Discovering

Number Theory

E 70

JOHN E. MAXFIELD

Professor of Mathematics
Kansas State University

and

MARGARET W. MAXFIELD

1972

W. B. SAUNDERS COMPANY • PHILADELPHIA • LONDON • TORONTO

W. B. Saunders Company: West Washington Square
Philadelphia, Pa. 19105

12 Dyott Street
London, WC1A 1DB

1835 Yonge Street
Toronto 7, Ontario

Discovering Number Theory

SBN 0-7216-6186-6

Print No: 9   8   7   6   5   4   3   2   1

Preface

We hope you will enjoy discovering number theory. Probably you will discover it in your own back yard, in the sense that many of your "discoveries" will be things you have known all along without giving them much thought. Number theory concerns the most familiar of all numbers, the numbers we count with, and it may surprise you to realize how much you already know about them from experience.

This book does not survey the whole field of number theory, but rather introduces you to many individual topics, some of which you may want to go into more thoroughly later. The first four chapters form a sequence, and Chapters 6 through 8 build on the ideas of Chapter 5. Otherwise the chapters are independent, although interrelationships are brought out. Hopefully there are enough topics included so that an instructor can vary the content of the course from time to time to keep the fresh attitude of discovery for each new class. In several chapters the last section or some of the later exercises can be skipped without loss of continuity.

Most of the chapters include directions for Suggested Projects: talks, demonstrations, or models suitable for mathematics clubs, science fair projects, or extra-credit reports. These will be especially useful to teachers.

Many thanks to our editor at W. B. Saunders Company, Carlos M. Puig, to the teachers who have taught us about numbers, and to our children, whose naive approach has helped us recall the joy of discovery.

JOHN E. MAXFIELD

MARGARET W. MAXFIELD

The Mathematics
of Amateurs

The mathematician G. H. Hardy was visiting his Indian protegé Ramanujan. Later Hardy wrote, "I had ridden in taxi-cab No. 1729, and remarked that the number seemed to me rather a dull one, and that I hoped it was not an unfavourable omen. 'No,' he replied, 'it is a very interesting number; it is the smallest number expressible as a sum of two cubes in two different ways.'"*

and also

$$1729 = 12^3 + 1^3 \text{ or } 1728 + 1$$

$$1729 = 9^3 + 10^3 = 729 + 1000.$$

Most of us have some natural curiosity about the "natural" numbers 1, 2, 3,... that we use for counting. At least they are more appealing than fractions! From very early days people have been interested in numbers, often attaching mystical importance to them. Some famous people have been attracted to numerology, including the composer J. S. Bach. Number theory is one branch of mathematics that has never been a fenced-off preserve for professional mathematicians. You can discover the subject without a whole new vocabulary of technical jargon.

You will find very little strain on your talents for arithmetic here. In fact, the few calculations are easy enough so that you can concentrate instead on satisfying your own curiosity. For example, you can learn how to make a simple slide rule, all with whole numbers less than 29, clarifying the whole idea of logarithms.

By the end of your study you will find that you have reviewed arithmetic quite painlessly, that you have learned something about axiomatic mathematics and deductive proofs, but most of all that you have revived your native excitement and curiosity about mathematics.

---

* See James R. Newman, "The World of Mathematics", Simon and Schuster, 1956, p. 375 for this anecdote, and pp. 366–376 for "Commentary on Srinivasa Ramanujan".

Contents

CHAPTER 1

The Integers

The "numbers" in "number theory" are the familiar ones you count with, so you will find it easy computationally. Since calculations are few and simple, you can concentrate on something much more central in mathematics—formulating and proving theorems.

You have already picked up a little number theory from observation: each multiple of 5 ends in 0 or 5; each even number ends in 0, 2, 4, 6, or 8; there are "prime" numbers that do not "factor;" and so on. You may have encountered other number theory ideas in puzzles and games. Although analytic techniques and even high-speed computers are used on some number theory problems, many others are accessible to any observant person. The field has been enriched by the contributions of many amateurs—ministers, businessmen, even professional gamblers, who have noticed facts about numbers, formulated conjectures, and substantiated their conjectures by proving theorems.

The counting numbers have been around a long time and we know quite a lot about them; yet there are still questions to be answered, some of them real puzzles that have baffled generations of mathematicians.

Definition 1.1 *The* **positive integers** *or* **natural numbers** *are the* **counting numbers**

$$1, 2, 3, 4, 5, \ldots.$$

Definition 1.2 *The* **integers** *are the counting numbers, their negatives, and zero:*

$$0, 1, -1, 2, -2, 3, -3, \ldots.$$

The non-negative integers are the positive integers and zero.

1.1 DISCUSSION OF THE COUNTING NUMBERS

In Definition 1.1 we do not try to define individual numbers and then define a collection of them; we define the whole system together with the way it works. This is like defining a ladder as a whole, with each individual rung as a contributor to the whole—the 4th rung, the 10th rung, and so on—not an object of attention by itself. The dots after the 5 are intended to suggest the whole progression of counting numbers, increasing by 1 at each step.

Sometimes people fear mathematics because it is "abstract." Perhaps they would feel more confident if they realized they have already mastered some important abstraction by just learning to count! They have learned to use the same counting system whether they are counting pictures or people, lions or lambs, books or bananas. They count with numbers that are themselves abstractions, for "3" is a concept abstracted from "3 stones," "3 years old," "3 bears," in fact, from all our experience with 3-membered sets.

Although Definition 1.1 lists five of the familiar arabic numerals, the definition is intended to apply to the *idea* of counting numbers, not just to the inked figures on the page. Conceptually, the definition would be exactly the same for Roman numerals I, II, III, IV, V, ..., or for a simple tally I, II, III, IIII, JHT, ...

How can we be certain we are all talking about the same idea, if that idea can be visualized in such different ways? Has any of us ever seen all the numbers written down, or heard them counted aloud above, say, 500? How do we know that if we started counting in some foreign language we might not diverge from English counting at some point? Since we are trying to describe an infinite system here, and consequently cannot write it all down and examine it, we give rules for generating it. One such rule that seems basic to the way we count sets is the mathematical induction principle, which we take to be a defining property of the positive integers.

The Mathematical Induction Principle: *Let S be a set of positive integers. Let S contain 1. For each positive integer s that S contains, let S also contain the next integer, $s+1$. Then S contains all the positive integers.*

This seems a plausible rule to adopt, for, since S contains 1, it contains the integer $s = 1$, and therefore contains $s+1$, or 2. Since it then contains $s = 2$, it must contain $s+1$ or 3. Since it then contains $s = 3$, it must contain $s+1 = 4$, and so on. This explanation does not *prove* the rule, for it involves the same kind of generation process from each s to the next that the rule provides for. The principle of mathematical induction is just stated to be a property of the counting numbers, one of the defining properties, so that if a set failed to exhibit the property we would say that it differed from the set of counting numbers.

Many proofs about integers use mathematical induction, since it is one of the few formal principles available about the basic process of counting. Here is an example of a proof by induction.

Proof by mathematical induction that every positive integer can be represented by tally marks:

Suppose you wanted to explain to someone how to keep a tally. Probably you would show him how to tally up to about 11 or 12.

I, II, III, IIII, ̶I̶I̶I̶I̶I̶, ̶I̶I̶I̶I̶I̶ I, ̶I̶I̶I̶I̶I̶ II, ̶I̶I̶I̶I̶I̶ III, ̶I̶I̶I̶I̶I̶ IIII, ̶I̶I̶I̶I̶I̶ ̶I̶I̶I̶I̶I̶, ̶I̶I̶I̶I̶I̶ ̶I̶I̶I̶I̶I̶ I, ̶I̶I̶I̶I̶I̶ ̶I̶I̶I̶I̶I̶ II, ...

Then you might say something like, "Each time you need to tally one more you add a mark to the bunch of marks you are working on, unless it has 4 or 5 in it. If it has 4 in it, you make a fifth mark across the 4. If it already has 5 marks, it is complete and you begin a new bunch." This description tells him all he needs to know to tally any counting number, because it shows how to generate the next tally. All we need to do to make the description into a proof by mathematical induction is to word it in terms of a set S.

Let S be the set of all positive integers that *can* be represented by tally marks. Then we are trying to prove that S contains all the positive integers. First, we establish what is called the **basis** of the induction by showing that S contains the number 1. As frequently happens, we establish this by demonstrating that 1 is in the set, here by showing the tally that represents it:

$$I \sim 1$$

Next, we must prove the **induction step**; that is, we must prove that, if s stands for a positive integer in S that can be represented by tally marks, then $s+1$ is also in S and can be represented by tally marks. As in the informal description, we tell how to generate the tally for $s+1$ from the tally for s by adding another mark. If the previous tally, that is the tally for s, ends with a bunch of 1, 2, or 3 marks, add a new mark to the bunch. If the last bunch has 4 marks, add a new mark across them. If the last bunch is complete with 5 marks, begin a new bunch. The resulting tally represents $s+1$, so S contains $s+1$. Therefore, by the mathematical induction principle, S contains all the positive integers.

1.2 ARITHMETIC

Addition and multiplication are defined on the counting numbers: addition in terms of counting, and multiplication in terms of addition. We use the positive integers to count the members in a set. If we count the members of two separate sets together in one count, without starting over at 1, the result is the **sum** of the two separate counts. Addition can be given a formal definition that shows how to generate sums involving $s+1$

($n + s + 1$, for instance) from sums involving s. Such a step-by-step definition is called "inductive," because it is keyed to the succession of the counting numbers, according to mathematical induction.

Multiplication can also be given a formal inductive definition based on adding, generating the product $n(s + 1)$ from the product $n \cdot s$.

Since the two basic operations of arithmetic are based on counting, their properties can be proved by the principle of mathematical induction. We shall merely list the properties here for reference purposes, but proofs may be found, for example, in E. Landau's *Foundations of Analysis*, Chelsea, 1960.

Definition 1.3 *For m, n, and r, any elements in an arithmetic with operations + and ×, we may have*

   i. associative laws: $(m + n) + r = m + (n + r)$

                                 $mn \cdot r = m \cdot nr$

 ii. commutative laws: $m + n = n + m$

                                 $mn = nm$

iii. cancellation laws: *If $m + r = n + r$, then $m = n$.*

                                 If $mr = nr$ and $r \neq 0$, then $m = n$.

 iv. distributive laws: $(m + n)r = mr + nr$

                                 $r(m + n) = rm + rn$

Definition 1.4 *Among the integers, 0 is the identity with respect to addition; that is, it is the integer for which*

$$n + 0 = 0 + n = n, \text{ for every integer } n;$$

for any positive integer m the corresponding negative $-m$ is the additive inverse, that is, the integer for which

$$m + (-m) = (-m) + m = 0 .$$

We define $0 \cdot m = m \cdot 0 = 0$, for each integer m.

Theorem 1.1 *The integers have the properties listed in Definition 1.3.*

Proof: The proofs are based on mathematical induction, hence on counting. We do not give the proofs here.

Sometimes we need to use the fact that the positive integers are ordered with a relation $>$, which can be defined in terms of counting or in terms of adding. We say $n > m$ (read "n is greater than m") if n comes after m in the sequence of counting. In terms of addition, we have

Definition 1.5   *Let m and n be positive integers. Then n > m if there is a positive integral difference d such that n = m + d.*

Points to Remember

1. The arithmetic of integers is based on the properties of counting. The rules or "laws" of arithmetic are neither arbitrary rules laid down by mathematicians or teachers, nor are they accidental discoveries; they are the natural outgrowth of the way we count and can be deduced systematically from counting.

2. The mathematical induction principle is not proved to be a property of the positive integers, but is postulated as a defining property. It gives a formal rule for generating the sequence of counting numbers.

SUGGESTED PROJECT

The properties of the positive integers and the idea of mathematical induction combine well with the notion of binary counting to make an attractive report, Mathematics Club talk, or science project. The material would be of special interest to an audience that had been working on change of base, but any audience prepared in junior high school mathematics can understand most of it, if it is presented mostly numerically with lots of examples.

We ordinarily record numbers in decimal notation, that is, to base 10, the position of each digit showing whether it counts units, tens, hundreds, thousands, and so on; for instance,

$$21435 \text{ means 2 ten thousands,}$$
$$\text{1 thousand,}$$
$$\text{4 hundreds,}$$
$$\text{3 tens,}$$
$$\text{and 5 units.}$$

However, occasionally we have use for numbers expressed to other bases, as in length measurements (base 12, 12 inches = 1 foot; or base 3, 3 feet = 1 yard) or in weight measurements (base 16, 16 ounces = 1 pound). Increasing international adoption of the metric system reflects a move to use base 10 consistently for measurements.

Binary, or base 2, notation has become fairly important because it adapts well to machine use, as we shall see. To write an integer in binary notation we need only the digits 0 and 1. To write 67 in binary notation, we find the highest power of 2 less than 67, which is $64 = 2^6$. From $67 = 64 + 3$, we have

$$67 = 2^6 + 2 + 1,$$
$$= 2^6 + 2^1 + 2^0$$

so in binary notation,

$$67_2 = 1000011.$$

To translate 98 into binary notation, we would write it in powers of 2: $98 = 64 + 32 + 2$, so that

$$98_2 = 1100010.$$

To translate from binary notation to decimal (base 10) notation, we determine from the position of each 1 which power of 2 it contributes. For instance, if

$$11010101$$

stands for a number written in binary notation, we know that the number is the sum of

one   1
one   4, or 2^2
one   16, or 2^4
one   64, or 2^6
and one   128, or 2^7
———
213

Then 11010101 translates to the base-10 number 213.

To write the first few positive integers in binary notation, imagine that you have a mechanical counter, like the mileage counter in a car, except that each dial has just the two binary digits 0 and 1, instead of the 10 decimal digits $0,1,2,3,4,5,6,7,8,9$. Before the count begins, all dials read "0". Now to count to 1, rotate the right dial, or "units" dial, to 1. To count to 2, rotate it again. As it changes from 1 to 0, it trips the next dial to its left, the 2's dial, analogous to the 10's dial in decimal counters. Then 2 written in binary notation is $2_2 = 10$, corresponding to $2 = 2^1 + 0 \cdot 1$. To count to 3, rotate the units dial to 1, getting $3_2 = 11$. To count to 4, rotate the units dial to 0, tripping the 2's dial. The 2's dial, in rotating from 1 to 0, trips the $2^2 = 4$'s dial, which rotates to 1. We have $4_2 = 100$, corresponding to $4 = 1 \cdot 2^2 + 0 \cdot 2 + 0 \cdot 1$.

The first 20 positive integers written in binary notation are

1	$6_2 =$ 110	1011	$16_2 =$ 10000
10	111	1100	10001
11	1000	1101	10010
100	1001	1110	10011
101	1010	1111	10100

Notice how each binary number in this list can be found from the previous one by rotating dials in the counter we described. Since each rotating dial has only the two digits 0 and 1, a switch with two settings would work just as well. It is this fact that makes binary counting just as natural for an electronic computer as decimal counting is for a person with 10 fingers. To make a simple demonstration of binary counting by switches, present a long row of playing cards, face down, or coins, all showing heads, or any other objects that have two definite sides that can serve as 0 and 1. For playing cards let the face down position stand for 0, so that before the count begins the display shows a long line of 0's. To count to 1, turn the right-most card face up. Then to count to 2, turn it face down and turn the second card from the right face up. Each time a card is turned face down, the next card to the left changes position.

Chapter 1 shows how to prove by mathematical induction that every positive integer can be represented by tally marks. The same kind of proof can show that every positive integer can be written in binary notation.

Let S be the set of all positive integers that can be written in binary notation. We have already established the basis of the induction, by showing how to write 1, and in fact, the first 20 numbers in binary notation. That is, we have shown that the first 20 numbers are in the set S. Now suppose some number s is in the set S. That means that it can be written in binary notation. We want to tell how to write $s+1$ in binary notation, but since we do not know exactly what number it stands for, we do not write its binary form, but give a definite rule for generating it. That rule is essentially the description of the way our mechanical counter advances. If the binary form of s ends in 0 in the units place, then change it to 1 and the result is the desired form for $s+1$. If the units digit of s is 1, then change it to 0 and change the next digit. If that next digit was 0 in s, it becomes 1 in $s+1$ and the increase is complete. For instance, if $s = 13$, the binary form for s would be 1101 and the binary form for $s+1$ would be 1110. At each stage if a digit changes from 0 to 1, the increase is complete and we have the binary notation for $s+1$. If the digit changes instead from 1 to 0, then the next digit to the left changes. The process stops, and the binary form for $s+1$ has been found, when either a 0 has been changed to a 1, or all the digits of s have been changed. Then $s+1$ is in S, since we have shown a sure technique for writing it in binary notation based on the binary notation for s. Then by mathematical induction, all the positive integers are in S, so that they can be written in binary notation.

Suppose someone doubts that there are big enough decimal forms to express very large positive integers. Can you adapt the proof by mathematical induction to convince him that every positive integer can be written in decimal notation?

CHAPTER 2

Divisors

We have learned from experience that an integer can be broken into irreducible factors. In this chapter we show that this factorization is unique (except for unit factors and order).

Figure 2.1 shows a typical factorization by trial and error. In order to prove results about the divisibility of integers, we study the division process itself, which we describe in Theorem 2.2.

Factor 14399.

By inspection, 2 is not a factor.

$$\begin{array}{r} 4799\ 2/3 \\ 3\overline{)14399} \end{array}\text{, so 3 is not a factor.}$$

By inspection, 5 is not a factor.

$$\begin{array}{r} 2057 \\ 7\overline{)14399} \end{array}\text{, so 7 is a factor, and } 14399 = 7\cdot 2057.$$

$$\begin{array}{r} 293\ 6/7 \\ 7\overline{)2057} \end{array}\text{, so 7 is not a factor of 2057.}$$

$$\begin{array}{r} 187 \\ 11\overline{)2057} \end{array}\text{, so that } 14399 = 7\cdot 2057 = 7\cdot 11\cdot 187.$$

$$\begin{array}{r} 17 \\ 11\overline{)187} \end{array}\text{, so that } 14399 = 7\cdot 11^2\cdot 17.$$

Figure 2.1 Factorization by trial and error.

EXERCISE 2.1. For this exercise use a bridge deck of 52 playing cards (or pebbles or blocks). First, deal all the cards to 2 players evenly. How many cards does each player have? Then we can write

$$52 = 2 \cdot \underline{\qquad} + 0.$$

Now redeal the 52 cards to 3 players, as many times as they will go evenly; you will need to keep one remainder card. Then

$$52 = 3 \cdot \underline{\qquad} + 1.$$

Now deal to 4 players.

$$52 = 4 \cdot \underline{\qquad} + \underline{\qquad}.$$

Show the quotient q and remainder r for 6 players, 7 players, 10 players, 13 players, and 14 players.

2.1 DIVISION

Before we can justify the division process in Theorem 2.2 we need to prove that for a positive dividend a, such as 52 in Exercise 2.1, and a positive divisor b, such as 5, there is always some quotient q large enough (10 in this case) so that the remainder r (here 2) is less than the divisor.

If the divisor b is a small positive integer and the dividend a is large, how can we be sure that there is a large enough q to keep the remainder small? As you will see, we use mathematical induction to prove it.

Theorem 2.1 *The positive integers have the Archimedean property; that is, given positive integers a and b, there is a positive integer n for which $bn > a$.*

Proof: We prove the theorem by mathematical induction. Let S be the set that contains each positive integer that *is* less than some multiple of b. We want to prove that S contains 1 and contains the next number $s+1$ for each of its members s. Then by mathematical induction we can conclude that S contains all the positive integers, in particular, a.

First, we establish the *basis*, that 1 is a member of the set S. Since b is a positive integer, we know from Definition 1.1 that b is in the sequence

Figure 2.2 For positive integers a and b, some multiple of b exceeds a.

1, 2, 3, ... Then $b \cdot 1 = b > 1$ if b is not 1, and $b \cdot 2 = 2 > 1$ if b is 1. This shows that there are sufficiently large multiples of b to surpass 1. Then 1 lies in the set S.

Next, we prove the *induction step*, that S contains $s + 1$ for each of its members s. Suppose that some s is a member of S. From the way we have defined the set S, this means there is a large enough multiple, say bn, of b so that $bn > s$. By Definition 1.5 there is a positive integer d for which $bn = s + d$. If $d > 1$, then $bn = (s + 1) + (d - 1)$, with $(d - 1)$ a positive integer, so that by Definition 1.5, $bn > s + 1$. In this case, then, the next number $s + 1$ is surpassed by bn and is therefore in S. What if $d = 1$? Then $bn = s + d = s + 1$, so that bn is not a large enough multiple of b to surpass $s + 1$. Try the next larger multiple of b.

$$b(n + 1) = bn + b = s + 1 + b.$$

Since $b > 0$, we have from Definition 1.5, $b(n + 1) > s + 1$. In this case, too, $s + 1$ lies in S. The induction is complete, so that all the positive integers, including the dividend a, have the property that they are surpassed by some multiple of b. ∎ [The symbol "∎" means that the proof is complete at this point. In some cases additional commentary may follow, but it is not part of the proof.]

EXERCISE 2.2.   Write 3 multiples of 4 that exceed 179.

EXERCISE 2.3.   Is there a multiple of 100 that exceeds 2?

Now we can introduce the division algorithm. An algorithm shows how a problem can be solved stepwise, thus proving constructively that it can be solved.

Theorem 2.2 [Division algorithm]   *Let b be a positive integer and let a be a non-negative integer. Then there is a non-negative integer q for which*

$$a = bq + r, \quad \text{with} \quad 0 \leqslant r < b.$$

Let a be 54, and let b be 8. Then a falls between two multiples of b,

$$0, 8, 16, 24, 32, 40, 48 \ (54) \ 56.$$

The remainder r is $54 - 48 = 6$, which is less than $b = 8$.

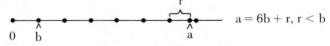

$$a = 6b + r, \ r < b$$

Figure 2.3   Division algorithm.

Proof: If $a = 0$, take $q = 0$ and $r = 0$. Since by Definition 1.6 we have $b \cdot 0 = 0$,

$$a = b \cdot 0 + 0 = 0 + 0 = 0.$$

If $a \neq 0$ and $a < b$, take $q = 0$ and $r = a$. Then

$$a = b \cdot 0 + a = 0 + a = a, \text{ with } 0 \leqslant a < b,$$

satisfying the requirements of the theorem.

Now suppose that $a \geqslant b$. We use the Archimedean property of Theorem 2.1 to conclude that there is a multiple bn of b for which $bn > a$. This multiple cannot be the first, $b \cdot 1$, because $a \geqslant b \cdot 1 = b$. Then somewhere in the finite sequence of multiples

$$b \cdot 1, b \cdot 2, b \cdot 3, \ldots, b(n-1), b \cdot n$$

is a multiple bq for which

$$bq \leqslant a < b(q+1).$$

Since $b(q+1) = bq + b$, we have

$$a = bq + r < bq + b,$$

so that $r < b$. ∎

EXERCISE 2.4. Find q and r as in Theorem 2.2 for these pairs a, b:

$$a = 17, \qquad b = 6$$
$$a = 5, \qquad b = 9$$
$$a = 1750, \qquad b = 25.$$

EXERCISE 2.5. Division is usually introduced as the inverse of multiplication ($12/3 = 4$, because $3 \cdot 4 = 12$, for instance), but show how someone who could not count above 3 could distribute 12 cocoanuts evenly among 3 people. Discover whether 7 divides 43 evenly by distributing pencil marks among 7 sets.

EXERCISE 2.6. Divide 16 into 260, getting the "wrong" quotient 15. What would the remainder be; that is, $260 = 16 \cdot 15 + ?$ Now get a different "wrong" quotient 17. What would the remainder be?

Discuss the restriction $0 \leqslant r < b$ in the division algorithm in the light of the "wrong quotient" idea, showing that the restriction is the familiar one we use to decide whether a trial quotient is "too small" or "too large."

EXERCISE 2.7.   A band of 16 non-counting monkeys try to divide 260 peanuts among themselves evenly, by dealing them around one at a time. Paralleling Exercise 2.6, describe the situation if they stop when each has 15 peanuts. Then explain what goes wrong if they try to deal a 17th peanut to each one. Use this exercise, Exercise 2.5, and Exercise 2.6 to relate the division algorithm to the practical problem of dividing a objects into b piles.

Definition 2.1 *Let m be an integer and d a non-zero integer. We call d a **divisor** of m if there is some integer q for which $m = qd$. If d is a divisor of m, we write $d \mid m$, read "d divides m"; if d is not a divisor of m, we write $d \nmid m$, read "d does not divide m."*

Notice that in Definition 2.1 there is a restriction on the divisor d. No attempt is made to define division by zero, because there would be ambiguity about the answer, or quotient. If we tried a divisor of $d = 0$ and the dividend m was not zero, there could be no quotient q for which $m = q \cdot 0$; on the other hand, if m was zero, then every integer would provide a quotient q.

Typically, definitions in mathematics are quite practical. Notice how we define divisor here in terms of the answer. We say 3 divides 123, because there is an answer for the division, and we say there is no meaning to the problem "Does 0 divide an integer?" because there would be either no answer or too many answers.

EXERCISE 2.8.   Show that $3 \mid 342$. Show that the problem is in the area covered by the definition, by mentioning that both 3 and 342 are integers and that $3 \neq 0$. What is q in this case? Is it an integer?

EXERCISE 2.9.   Show that $2 \nmid 7$. Would there be a quotient q if we did not restrict ourselves to integers?

EXERCISE 2.10.   Does 3 divide 0? Is the q for this exercise an integer?

EXERCISE 2.11.   Does -3 divide 342? $117 \mid 0$? $-7 \mid 0$?

EXERCISE 2.12.   We can show that 0 is divisible by integers greater than itself. Is this true of any other integers? (Try -12.) Is it true of any positive integers?

EXERCISE 2.13.   Is every non-zero integer divisible by itself? (Why the restriction "non-zero"?) If so, can you give a formula for finding q in this case? Is every integer divisible by 1? What is the formula for q?

Figure 2.4   Divisor d of m.

EXERCISE 2.14.   Is every integer m divisible by integers other than $+m$, $-m$, $+1$, and -1? (If you want to prove some conjecture false, all you need do is show one case in which it fails. If you want to show here that not every integer has other divisors, you need only exhibit one that has not.)

EXERCISE 2.15.   Write 4 different multiples of 6. Letting m stand for any integer, show how to write algebraically "any multiple of 6." Show how to write "any multiple of 11." Show how to write any even integer.

EXERCISE 2.16.   Letting m stand for any integer, write the algebraic expression for any even integer and for any multiple of 4. Prove that if an integer is a multiple of 4, then it must be even.

EXAMPLE.   Prove that if $a \mid b$ and $a \mid c$, then $a \mid (b-c)$.

By Definition 2.1, $a \mid b$ means that there is a quotient q for which $b = aq$, and $a \mid c$ means that there is a quotient k for which $c = ak$. Then $b - c = aq - ak = a(q-k)$. Let j stand for the integer $(q-k)$. Then $b - c = a(q-k) = aj$, so that by Definition 2.1, $a \mid (b-c)$, with quotient j. Note that a is not zero, since it is given as a divisor of b and of c. If j is zero, then b and c are equal, and a divides $(b-c) = 0$.

EXERCISE 2.17.   Prove that if $a \mid b$, then $a \mid bx$ for each integer x.

EXERCISE 2.18.   Prove that if $a \mid b$ and $a \mid x$, then $a \mid (b+x)$.

EXERCISE 2.19.   Prove that if $a \mid b$ and $a \mid x$, then $a \mid (rb+sx)$, where r and s are integers.

EXERCISE 2.20.   Prove that if $a \mid b$ and $b \mid x$, then $a \mid x$. (This is called the "transitive" property of division.)

2.2   GREATEST COMMON DIVISOR

Suppose a restaurant has a standing order for 18 packages of frozen filets at each delivery, while a smaller restaurant has an order for 12 packages. Then if the packages come in bunches of 2 tied together, the delivery man can leave 9 bunches for the first restaurant and 6 for the smaller. If the packages are in bunches of 3, the larger restaurant needs 6 bunches and the smaller needs 4. What is the largest bunch of packages that could be used to serve these two customers without splitting a bunch? Your experience with numbers in just such a context as this tells you that if the packages are tied in bunches of 6 they will be the handiest possible for these two deliveries. All the desirable sizes for bunches in this case, 2, 3, and 6, are what we call "common divisors" of 12 and 18; we call 6 the "greatest common divisor." (As we shall see, 1, -1, -2, -3, and -6 are also common divisors.)

Definition 2.2 *Let a and b be integers, not both zero. An integer d is a* **common divisor** *of a and b if $d\,|\,a$ and $d\,|\,b$.*

EXERCISE 2.21.   Show that the common divisors of 18 and 30 are 1, -1, 2, -2, 3, -3, 6, and -6. Save your work to use in Exercise 2.23.

EXERCISE 2.22.   Let a and b be integers, not both zero. Prove that 0 is not a common divisor of a and b.

Definition 2.3 *Let a and b be integers, not both zero. An integer g is a* **greatest common divisor**, *or* **GCD**, *of a and b if*
  i. *g is a common divisor of a and b, and*
  ii. *g is divisible by every common divisor of a and b.*
We often list only the positive common divisors and greatest common divisor, abbreviating the positive GCD of a and b as $(\mathbf{a}, \mathbf{b})$.

EXAMPLE.   Let a be 24 and let b be 60. Then the common divisors of a and b are 1, 2, 3, 4, 6, 12, and their negatives, -1, -2, -3, -4, -6, and -12. The greatest common divisors are 12 and -12, since they are common divisors that are divisible by every common divisor. We have $(a, b) = 12$.

EXERCISE 2.23.   Using the result of Exercise 2.21, show that 6 and -6 are greatest common divisors of 18 and 30.

EXERCISE 2.24.   Find the positive GCD of 12 and 15, or $(12, 15)$. Find the positive common divisors and verify that they all divide $(12, 15)$.

EXERCISE 2.25.   Find $(52, 117)$ and how that all the negative common divisors divide $(52, 117)$.

In Theorem 2.3 we will develop a systematic way to find (a, b) without writing all the common divisors of a and b. First, in order to sharpen intuition, we introduce an example from experience with measuring lengths. As we know from experience, if d and m are positive lengths, then $d\,|\,m$ means that m can be "measured" in d's; that is, a stick m units long can be measured by several consecutive placings of a d-unit ruler with no remainder, as shown in Figure 2.5.

In terms of measurements, a common divisor of two lengths, say 52 units and 14 units, is a length d that can be used to measure both 52 and 14. The greatest common divisor is the longest such measure.

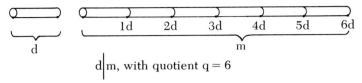

$d\,|\,m$, with quotient $q = 6$

Figure 2.5 If $d\,|\,m$ with quotient q, then a stick m units long can be measured by q placings of a stick d units long.

Suppose we start with two unmarked sticks, and we want to find the longest length d that will measure both of them. For definiteness we suppose here that the two unknown lengths are 52 and 14 units. We have no ruler to measure with, just the two sticks. We can see by comparison that one stick is longer than the other, so a natural start on finding d is to use the shorter stick to measure the longer one as closely as we can. Accordingly, we break as many 14-unit lengths as possible from the 52-unit stick using the shorter stick as measure. When we have removed three 14-unit lengths from the 52-unit stick we will have a 10-unit stick left over as remainder, a new measure shorter than the 14-unit stick. (See Figure 2.6.)

Again, it seems natural to try to measure the 14-unit stick by finding how many 10-unit lengths it contains. We find it contains one, with a remainder 4 units long (Fig. 2.7).

Again, we measure the 10-unit stick with the 4-unit stick, getting two, plus a 2-unit remainder. (Fig. 2.8).

At the next stage we reach an end to the process, for the 2-unit stick will measure the 4-unit stick exactly, in two placings. (Fig. 2.9). Also, we shall see that we have reduced our homemade measuring sticks until we have the longest common measure we can use to compare 14 and 52. That is, we can express 14 in 2's, as $2 \cdot 7$, and 52 in 2's, as $2 \cdot 26$, and we would not be able to express both in 3's, 4's, 5's, or any bigger integer.

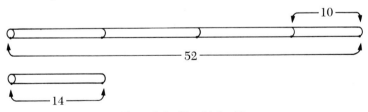

Figure 2.6 $52 = 14 \cdot 3 + 10$.

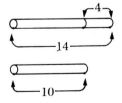

Figure 2.7 $14 = 10 \cdot 1 + 4$.

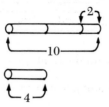

Figure 2.8 $10 = 4 \cdot 2 + 2$.

Figure 2.9 $4 = 2 \cdot 2 + 0$.

We can prove that this scheme works in general. We need only replace 52 and 14 in the above example by a general a and b and justify each division by Theorem 2.2:

Theorem 2.3 [Euclidean algorithm]   *Let a and b be positive integers. Then (a, b) can be found in a finite number of steps as a linear combination of a and b:*

$$(a, b) = ra + sb, \qquad \text{where } r \text{ and } s \text{ are integers.}$$

Proof: [Follow Figure 2.10 for a special case.] By Theorem 2.2 there is a non-negative quotient q_1 and a remainder r_1 for which

(1) $$a = bq_1 + r_1, \text{ and } 0 \leqslant r_1 < b.$$

If $r_1 \neq 0$, then we may use Theorem 2.2 to divide b by r_1:

$$b = r_1 q_2 + r_2, \text{ with } 0 \leqslant r_2 < r_1.$$

If $r_2 \neq 0$, we use Theorem 2.2 to divide r_1 by r_2:

$$r_1 = r_2 q_3 + r_3, \text{ with } 0 \leqslant r_3 < r_2.$$

We continue in this way. So long as the latest remainder is greater than zero, we divide it into the previous remainder. However, notice that the remainders decrease at each step. Therefore, since the first remainder r_1 was less than b, in a finite number of steps we must find some remainder, say r_e, that equals zero. We will show that $r_{e-1} = (a, b)$. [If $e = 1$, take $b = r_0$.] From the division step with remainder r_e,

(e)
$$r_{e-2} = r_{e-1}q_e + r_e, \text{ with } r_e = 0,$$

we have $r_{e-1} \mid r_{e-2}$. But from the previous division

(e − 1)
$$r_{e-3} = r_{e-2}q_{e-1} + r_{e-1},$$

so $r_{e-1} \mid r_{e-3}$, also. Continuing backward through the division steps, we can show that r_{e-1} divides each remainder and then that it divides b and a. Then it is a common divisor of a and b.

We can express r_{e-1} in the linear form mentioned in the theorem, $ra + sb$, for Equation (e − 1) gives us a way to write r_{e-1} in terms of two previous remainders as

$$r_{e-1} = r_{e-3} - r_{e-2}q_{e-1};$$

from Equation (e − 2) we can write r_{e-2} in terms of the two previous remainders as

$$r_{e-2} = r_{e-4} - r_{e-3}q_{e-2},$$

and so on, with r_2 finally expressed in terms of r_1 and b,

$$r_2 = b - r_1q_2,$$

and r_1 expressed in terms of a and b, from Equation (1):

$$r_1 = a - bq_1.$$

Collecting terms after performing all these substitutions gives us a form

$$r_{e-1} = ra + sb.$$

This provides the form required in the theorem and at the same time enables us to prove requirement *ii* in Definition 2.3, for by the distributive law any common divisor of a and b can be factored out of $ra + sb$ and hence divides r_{e-1}. Then $r_{e-1} = (a, b)$. ∎

EXERCISE 2.26.   In Exercise 2.23 you found $(52, 117)$. Express it in the form $r \cdot 52 + s \cdot 117$.

EXERCISE 2.27.   Find $(18, 105)$ and express it linearly in terms of 18 and 105.

EXERCISE 2.28.   Find $(16, 152)$ and express it linearly in terms of 16 and 152.

EXERCISE 2.29.   Find $(341, 661)$ by the Euclidean algorithm.

$$(1) \quad 52 = 14 \cdot 3 + 10$$

$$(2) \quad 14 = 10 \cdot 1 + 4$$

$$(3) \quad 10 = 4 \cdot 2 + 2$$

$$(4) \quad 4 = 2 \cdot 2 + 0$$

From (4), $2 \mid 4$. From (3), then,

$$2 \mid 10 \quad \text{which equals} \quad 2[2 \cdot 2 + 1].$$

From (2), then,

$$2 \mid 14 \quad \text{which equals} \quad 2[5 \cdot 1 + 2].$$

From (1), then,

$$2 \mid 52 \quad \text{which equals} \quad 2[7 \cdot 3 + 5].$$

Then 2 is a common divisor of 52 and 14.

Now to obtain 2 as a linear form in 52 and 14, use (3) to write

$$2 = 10 - 4 \cdot 2.$$

Then substitute for 4 the linear form obtainable from (2):

$$4 = 14 - 10 \cdot 1,$$

giving

$$2 = 10 - [14 - 10 \cdot 1]2.$$

But from (1), 10 can be expressed linearly as

$$10 = 52 - 14 \cdot 3,$$

giving us for 2

$$2 = [52 - 14 \cdot 3] - [14 - (52 - 14 \cdot 3)1]2.$$

Collecting terms in 52 and in 14, we have the expression

$$2 = (3)(52) + (-11)(14).$$

Figure 2.10   The Euclidean algorithm for the case $a = 52$, $b = 14$.

EXERCISE 2.30.   Find $(177, 252)$ and express it linearly in terms of 177 and 252.

EXERCISE 2.31.   Extend Theorem 2.3 by showing how to express $(0, b)$ linearly in terms of 0 and b, where $b > 0$.

EXERCISE 2.32.   Extend Theorem 2.3 by finding formulas for $(-a, b)$, $(a, -b)$, and $(-a, -b)$, where a and b are positive integers. Treat the case $(0, -b)$. State a more general result than Theorem 2.3 justified by Exercise 2.31 and Exercise 2.32.

2.3   THE IDEAL (a, b)

We found r and s as a by-product of a process or algorithm for finding (a, b). Now we take a completely different approach, which will help us interpret what we have done. Think of all the different integers n that can be written in the form

$$(1) \qquad\qquad n = ax + by.$$

One such integer is (a, b), which we get when we use r in the x position and s in the y position. In fact, any multiple $m(a, b)$ is such an integer, with x as mr and y as ms. Suppose d is any common divisor of a and b, so that $a = da'$ and $b = db'$ for some a', b'. Then $n = da'x + db'y = d(a'x + b'y)$, so $d \mid n$. Thus any integer n that can be written in the form $ax + by$ is a multiple of (a, b). Since (a, b) is a common divisor of a and b, we have $(a, b) \mid n$. Then the integers n we are looking for in (1) are all the integral multiples of (a, b), and (a, b) has a minimal property among all the integers n in (1); namely, it is the smallest positive one.

All the integers in (1) form what algebraists call an **ideal** among the integers: it is a set of integers selected so that if two of them are added or subtracted the answer is again one of the integers in (1), and so that if one of them is multiplied by any integer the answer is also one of the integers in (1).

Now (1) can be graphed on cartesian coordinates as a straight line with slope $-a/b$ and y-intercept n/b (Fig. 2.11). As n ranges over the multiples of

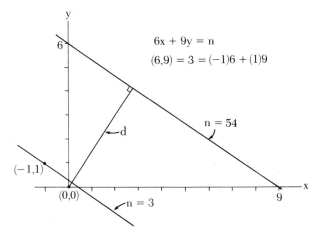

Figure 2.11   The ideal of integers $n = 6x + 9y$ can be represented graphically as a family of parallel lines.

(a, b) the graph ranges over a family of parallel lines. It can be shown that the perpendicular distance d between any of the lines and the origin $(0,0)$ is $n/(a^2 + b^2)$. Since (a, b) is the smallest positive n in (1), the line of the family $ax + by = n$ corresponding to (a, b) is the one with smallest n, and hence is the line nearest $(0,0)$ that passes through a lattice point (r, s). [A lattice point is a point whose coordinates are integers, one of the intersections on the cartesian grid.]

Points to Remember

1. So long as an integer b is not zero, we can divide by it in such a way that the remainder is less than b. If b is positive and a is non-negative, then $a = bq + r$, with $0 \leqslant r < b$. Division by 0 is not defined.

2. Two integers not both zero have a greatest common divisor, and there is a systematic process called the Euclidean algorithm for finding it and expressing it linearly in terms of the two integers.

3. The greatest common divisor of two integers a and b is the smallest positive integer among the linear expressions $ax + by$, where x and y are integers.

SUGGESTED PROJECTS

1. Figures 2.5 through 2.9 suggest a display of physical models that could be accompanied by a talk to explain the Euclidean algorithm. The whole nature of division is clarified for some by the notion of measurement. If there is need to keep the talk as elementary as possible, restrict it to numerical examples to accompany the models.

2. Figure 2.11 can be adapted to make a very attractive physical model. The x and y axes can be drawn on large-scale graph paper, and the line $6x + 9y = n$ on a transparent sheet, pliofilm, or celluloid. Arrange the display so that the transparent sheet can be moved across the graph paper with its line always at the proper slant. Illustrate various settings and show that the lattice points (intersections on the graph paper) that fall on the line give $6r + 9s = n$. Show which placing of the transparent sheet yields $(6, 9) = 3$.

CHAPTER 3

Primes

3.1 RELATIVELY PRIME INTEGERS AND PRIME INTEGERS

When we introduced common divisors, we used an example about two restaurants, one needing 18 packages of filets, the other 12, and we noticed that deliveries could be made in bunches of 2, or in bunches of 3, or most conveniently in bunches of 6. Now suppose the smaller restaurant increases its order to 17 packages per delivery. Then the delivery man must use single packages as units or else split bunches. In fact, 17 has very special properties as numbers go. No matter what other restaurants he has on his list, the delivery man cannot use any convenient bunching arrangement except for customers needing multiples of 17, such as 34 or 51. We say that 18 and 17 are "relatively prime" and that 17 is a "prime integer" or a "prime."

Definition 3.1 *An integer a is* **relatively prime** *to an integer b if whenever it divides an integral multiple bn of b then it divides n; that is, $a \mid bn \rightarrow a \mid n$. (The arrow is read "implies.")*

Theorem 3.1 *An integer a is relatively prime to b if and only if $(a, b) = 1$.*

Proof: First, to prove the "only if" part, let a be relatively prime to b and let $(a, b) = d$. Since $d \mid a$ and $d \mid b$, we can write $a = a'd$ and $b = db'$, from Definition 2.2. Then $a \mid ab'$, where $ab' = (a'd)b' = a'(db') = a'b$. Then $a \mid a'$, because a is relatively prime to b. But $a = a'd$, so $d = 1$.

Conversely, to prove the "if" part, suppose that $(a, b) = 1$. From Theorem 2.3 we can write

$$(1) \qquad\qquad 1 = ra + sb,$$

for some integers r and s. Now suppose that $a \mid bn$ for some multiple bn of b. Multiply (1) by n to get

$$n = ran + sbn.$$

Then $a \mid ran$ and $a \mid sbn$, so $a \mid n$, where $n = ran + sbn$. (See Exercise 2.18.) Thus, by Definition 3.1, a is relatively prime to b. ∎

EXAMPLE.   Since $(6, 7) = 1$, any multiple $7n$ of 7 that is divisible by 6 must have $6 \mid n$:

We have $6 \nmid 7 \cdot 1$, $6 \nmid 7 \cdot 2$, $6 \nmid 7 \cdot 3$, ..., $6 \nmid 7 \cdot 5$, $6 \nmid 7 \cdot 21$, ..., but $6 \mid 7 \cdot 6$, $6 \mid 7 \cdot 18$, $6 \mid (-7) \cdot 24$, ...

Since $(6, 10) = 2$, there are multiples $10n$ of 10 that are divisible by 6 with n not divisible by 6. For instance, $6 \mid 10 \cdot 3$, $6 \mid 10 \cdot 9$, $6 \mid 10 \cdot 15$, and $6 \mid 10 \cdot 3k$.

EXERCISE 3.1.   Let $d = (a, b)$ and let $a = da'$, $b = db'$. Prove by Theorem 3.1 that a' and b' are relatively prime.

EXERCISE 3.2.   Find two numbers, each greater than 1000, that are relatively prime.

EXERCISE 3.3.   Are 177 and 148 relatively prime?

EXERCISE 3.4.   Show that 61 is relatively prime to all integers greater than one except multiples of 61.

Definition 3.2   *An integer p is a* **prime** *if $p > 1$ and if whenever it divides a product bc of integers then it divides one of the factors b or c; that is, $p \mid bc \to p \mid b$ or $p \mid c$.*

Definition 3.3   *An integer p is* **irreducible** *if $p > 1$ and if p has no divisors strictly between 1 and p; that is, no $d \mid p$ with $1 \leqslant d < p \to d = 1$.*

EXAMPLE.   If the prime 3 divides a product such as $42 = 6 \cdot 7$, then either $3 \mid 6$ or $3 \mid 7$.

Since 6 is not a prime, 6 can divide a product such as $10 \cdot 9 = 90$ without dividing either factor:

$$6 \nmid 10 \text{ and } 6 \nmid 9.$$

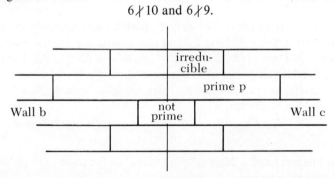

Figure 3.1   If a prime block p contributes to a two-part wall, bc, then it lies entirely in b or entirely in c. An irreducible block cannot be broken into smaller blocks.

Theorem 3.2   *An integer p is prime if and only if it is irreducible.*

Proof: (Figure 3.1 may help you to relate the two concepts.) Suppose that p is a prime. If p has a divisor d, then $p = dp'$ for some positive integer p'. Since p divides itself, $p \mid p$, p is a divisor of the product dp'. Then since p is a prime it must divide one of the factors, $p \mid d$ or $p \mid p'$. But $p \geqslant d$ and $p \geqslant p'$, so the only divisors are ± 1 and $\pm p$.

Conversely, suppose that p is irreducible. Then its positive divisors are 1 and p. Then its positive common divisors with any integer b are 1 and, in case $p \mid b$, p. Suppose that $p \mid bc$. Then either $p \mid b$ or $(p, b) = 1$. In the latter case, by Theorem 2.3, there are integers r and s for which

$$1 = rp + sb.$$

Multiplying by c, we have

$$c = rpc + sbc.$$

Since $p \mid rpc$ and $p \mid bc$, we have $p \mid c$, where $c = rpc + sbc$. This proves that if $p \mid bc$, then either $p \mid b$ or $p \mid c$; that is, p is a prime. ∎

Both Theorem 3.1 and Theorem 3.2 are "if and only if" results. They give "necessary and sufficient conditions" for an integer to be relatively prime to another integer or to be a prime integer, respectively. Any condition that is necessary and sufficient to a definition, such as the definition of "prime," characterizes the object completely and so could itself be used as a definition. Thus it is arbitrary whether we define "prime" as in Definition 3.2 or as an integer > 1 that has only itself and 1 as positive divisors (Definition 3.3), and we often see primes defined in the latter way.

An advantage to having a separate word for the idea "irreducible" is that there are systems in which some elements are irreducible but are not primes. The fact that among the integers the irreducible elements coincide exactly with the primes, as proved in Theorem 3.2, is basic to our proof of unique factorization for integers.

Theorems 3.1 and 3.2 make it possible to check for relatively prime integers and prime integers by looking at their divisors instead of looking at their infinite number of multiples, as would be required by Definitions 3.1 and 3.2. For instance, to see that 101 is a prime we need not look at all products bc divisible by 101 to be sure that $101 \mid b$ or $101 \mid c$. We can simply check to see whether 101 has any divisors between 1 and 101.

EXERCISE 3.5.   To show that 101 is a prime we need not look at negative divisors. Why?

EXERCISE 3.6.   If 101 had a divisor less than itself and greater than 10, how big would the quotient be? Use this idea to show that in checking for divisors of an integer n we do not need to try divisors $> \sqrt{n}$.

EXERCISE 3.7   Suppose that $d \nmid n$. Then show that $dm \nmid n$. [Suppose some multiple dm of d does divide n. Let the quotient be q. Show that this would imply $d \mid n$, contrary to the assumption.]

Definition 3.4   *Integers > 1 (or < -1) that are not primes are called* **composites**. *The numbers 1 and -1 are* **units**.

3.2   THE SIEVE OF ERATOSTHENES

One way to find primes is to wash away the composite integers, leaving the primes caught in the "Sieve of Eratosthenes." To illustrate the method, list the positive integers from 2 to 100. As in Figure 3.2, circle 2 as the first prime in the list, but strike out all larger multiples of 2. Circle 3, but strike out all larger multiples of 3. (Notice that alternate multiples of 3 have already been washed through the sieve as multiples of 2.)

EXÉRCISE 3.8.   Continue this process, finding 25 positive primes < 100.

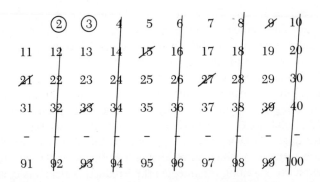

Figure 3.2   The Sieve of Eratosthenes.

Theorem 3.3   *If an integer > 1 is not a prime, then it is divisible by a prime.*

Proof: By Theorem 3.2, an integer n greater than 1 that is not a prime has some factorization $n = n_1 n_2$ for which $1 < n_1 < n$. Among all such factorizations, choose one with n_1 minimal. Then n_1 is a prime, for if not, then $n_1 = n_{11} n_{12}$ with $1 < n_{11} < n_1$, so that $n = n_1 n_2 = (n_{11} n_{12}) n_2 = n_{11}(n_{12} n_2)$. This contradicts the assumption that n_1 is the smallest divisor of n. ∎

We can apply this to prove Euclid's famous theorem that there are infinitely many primes.

Theorem 3.4   *There are infinitely many primes.*

Proof: Suppose that there are only a finite number of primes, $p_1, p_2, ..., p_n$, so that p_n is the largest prime. Form the number $m = (p_1 p_2 \cdots p_n) + 1$, the product of all the primes plus 1. The number m is an integer, since it is formed by multiplying and adding integers, and m is greater than 1. Then by Theorem 3.3 either m itself is a prime or it is divisible by a prime. But from $m - (p_1 p_2 \cdots p_n) = 1$, m cannot be divisible by any of the primes in the known list, $p_1, p_2, ..., p_n$, for if p_i, say, divided m, then it would have to divide 1. Then either m or some divisor of $m \neq p_i$ is a prime, contrary to the supposition that the finite list contained all the primes. ∎

The primes have had a special interest for people since early days. Most of the "lucky" and "unlucky" numbers are primes: "Catastrophes come in threes," "The third time is the charm," "Seven come eleven," "Unlucky 13," and so on.

As you saw in Theorem 3.4, there are infinitely many primes, so there are arbitrarily large primes. (See Exercise 3.9.)

EXERCISE 3.9. Substantiate the preceding sentence in this way: Suppose there were no primes larger than some large prime P. Then prove that the total number of primes would be a finite number less than P, contrary to Theorem 3.4.

Some very large prime numbers are known, such as $2^{11213} - 1$, which if expanded would have 3376 digits.

What about the density of primes? We know from Theorem 3.4 and Exercise 3.9 that there are primes arbitrarily far out in the sequence of positive integers, but does the frequency of primes, that is, the ratio of the number of primes less than or equal to n, called $p(n)$, to n increase among very large integers? No, it has been proved in the *Prime Number Theorem* that the limit

$$\lim_{n \to \infty} \frac{p(n) \ln n}{n} = 1.$$

The terms "limit" and "ln" (for "natural logarithm") arise in calculus. For our purposes it suffices to say that, for large enough n, $p(n)(\ln n)/n$ is arbitrarily near 1.

EXERCISE 3.10. Find ln 100 in a table of natural logarithms (logarithms to base e). Compare $p(100)/100$ with $1/\ln 100$.

The Prime Number Theorem was not proved for a century after it was guessed from observation. As this history suggests, the proof is not easy.

Primes still provide tantalizing puzzles. We have no formula to produce the primes. We have turned computers loose on the prime hunt, mostly using modifications of the ancient Sieve of Eratosthenes, but we have no way to crank all the primes out, to construct them according to some rule.

Another famous problem is that of "twin primes." The pairs 3 and 5, 5 and 7, 11 and 13, 17 and 19 are sets of twin primes; that is, primes that differ by 2.

EXERCISE 3.11.   Find a pair of primes that differ by 1 and prove that they are the only such pair.

The famous question is, are there infinitely many twin prime-pairs? D. H. and E. Lehmer* found that there are 152,892 pairs of twin primes less than 30,000,000. A large known pair is $10^{12} + 9650 \pm 1$.

Points to Remember

1. Primes are the multiplicative building blocks of the integers.

2. It has been known since Euclid that there are infinitely many primes and we now know from the Prime Number Theorem how they are distributed, at least in a general way. However, many interesting questions about primes are still unanswered.

SUGGESTED PROJECTS

Nowhere is the layman's natural interest in numbers more prominent than in questions about primes. Fortunately, some of the most intriguing possibilities can be described with very little preparation. For an elementary talk it is best to treat prime numbers simply as irreducible integers, since they do coincide, as Theorem 3.2 shows.

After a brief description of primes, composites, and units, with examples, a good subject is the Sieve of Eratosthenes. Many of the greatest mathematicians spent their spare time on "prime hunts," and today we have armies of computers available to look for primes. Most of the computer techniques are based on adaptations of the Sieve method.

Most audiences can grasp the ideas in Theorems 3.3 and 3.4. Euclid's theorem (3.4) makes an especially nice topic for an elementary audience, as it is relatively simple, yet important and characteristic.

The many unsolved problems about primes tend to capture interest and can be stated easily without introducing a lot of vocabulary.

---

* "Density of primes having various specified properties," Communication to Section IIB of the Inter. Congress of Math. (Edinburgh, 1958)

CHAPTER 4

Unique Factorization

4.1 THE FUNDAMENTAL THEOREM OF ARITHMETIC

For many purposes integers are easiest to use when they are broken down into irreducible factors. By repeated application of Theorem 3.2 we can factor any integer > 1 into a product of primes. If the given integer is a prime, we still call it a "product" of primes.

EXAMPLE. $63 = 3 \cdot 3 \cdot 7 = 3^2 \cdot 7$

$141,192 = 2^3 \cdot 3^2 \cdot 37 \cdot 53$

$487 = 1 \cdot 487$, with no integral divisors between 1 and 487.

With certain conventions we can show that not only does each integer have a prime factorization, but that it has *only* one; that is, the factorization is unique. The conventions concern unit factors (since an arbitrary number of these might multiply the product of primes) and the order of the prime factors. We decide arbitrarily to list the positive prime factors in increasing order of size (adjoining primes may be equal) and to write down no unit factors unless the given integer is negative, in which case we include one -1.

EXERCISE 4.1.   Factor 726, -1431, and 4334 into primes, following the conventions of the preceding paragraph.

Theorem 4.1 [The Fundamental Theorem of Arithmetic]   *Every integer > 1 (or < -1) has exactly one factorization into primes, with the conventions above.*

Proof: Suppose n is an integer greater than 1. Then by Theorem 3.3 either n is a prime, or it is divisible by a prime p, say $n = pq$, for some quotient q. If $n = pq$, another application of Theorem 3.3 shows that either q is a prime or it is divisible by a prime, say $q = p_2 q_2$. Again by Theorem 3.3, either q_2 is a prime or $q_2 = p_3 q_3$, where p_3 is a prime. In this way we can show by repeated applications of Theorem 3.3 that any integer $n > 1$ (or < -1) has some factorization into a product of primes.

Suppose some integer $n > 1$ has two different factorizations, say

$$n = p_1 p_2 \cdots p_r \quad \text{and} \quad n = q_1 q_2 \cdots q_s,$$

with the primes listed in increasing order of size, $p_i \leqslant p_j$ for $i < j$, and similarly for the q's. Since p_1 divides n in its p-factorization, $n = p_1 p_2 \cdots p_r$, then p_1 must also divide n in its q-factorization:

$$p_1 \mid n, \quad \text{where } n = q_1 q_2 \cdots q_s.$$

Does $p_1 \mid q_1$? Only if $p_1 = q_1$. If $p_1 \neq q_1$, then $p_1 \mid q_2 \cdots q_s$, by Definition 3.2. Similarly, either $p_1 = q_2$, or $p_1 \mid q_3 \cdots q_s$, and so on. Eventually p_1 must equal one of the q's. Then $p_1 \geqslant q_1$, the smallest of the q's. Now treat q_1 similarly, showing that q_1 equals one of the p's and $q_1 \geqslant p_1$. Then $p_1 = q_1$.

We divide n by $p_1 = q_1$, getting

$$n/p_1 = p_2 p_3 \cdots p_r = q_2 q_3 \cdots q_s.$$

As before, we can show that $p_2 \geqslant q_2$ and $q_2 \geqslant p_2$, so that $p_2 = q_2$; and so on.

We need to prove that the number of primes is the same in both factorizations. Suppose this were not so. If $s > r$, say, we would have after r steps

$$n/p_1 p_2 \cdots p_r = 1 = q_{r+1} q_{r+2} \cdots q_s,$$

but each q is a prime, and so is greater than 1. Then $r = s$ and the "two" factorizations are the same. ∎

EXERCISE 4.2.   The integer 9333 has 61 as a factor. Start with this divisor and find the other factors. However, notice that 3 divides 9333. Starting with 3, factor 9333 in a different order. Compare these two factorizations.

You will appreciate Theorem 4.1 more fully when you study Fermat's Last Theorem.

4.2 GCD AND LCM OF FACTORED INTEGERS

Once we have two integers a and b factored into prime powers we can find (a, b) very easily:

Theorem 4.2   *Let $a = p_1^{a_1} p_2^{a_2} \ldots p_r^{a_r}$ and $b = p_1^{b_1} p_2^{b_2} \ldots p_r^{b_r}$, where the p's are primes and the exponents a_i, b_i are non-negative integers. Then*

$$(a, b) = p_1^{c_1} p_2^{c_2} \ldots p_r^{c_r},$$

where each $c_i = min\{a_i, b_i\}$.

Proof: Notice what the notation of the theorem means. Ordinarily a and b will not be divisible by exactly the same primes. We list all r primes that divide either one or the other (or both), letting $a_2 = 0$, say, if p_2 fails to divide a, with the convention $p^0 = 1$. For an example, see Figure 4.1.

The divisors of a are the integers of the form $d = p_1^{d_1} p_2^{d_2} \ldots p_r^{d_r}$, where $0 \leqslant d_i \leqslant a_i$ for each i, because each such d divides a, as we can show by giving the factorization

$$a = d(p_1^{a_1 - d_1} p_2^{a_2 - d_2} \ldots p_r^{a_r - d_r}),$$

and no other integers divide a, since $p_i^{e_i} \nmid a$ if $e_i > a_i$ or if p_i is not one of the primes in the factorization of a. From this observation,

$$p_1^{c_1} p_2^{c_2} \ldots p_r^{c_r}$$

divides both a and b and is in turn divided by every common divisor. ∎

Corollary 4.1   *Let $d = (a, b)$, $a = a'd$, $b = b'd$. Then $(a', b') = 1$.*

Let $a = 2^3 \cdot 3^2 \cdot 7$ and $b = 2 \cdot 5 \cdot 7 \cdot 17$. To find (a, b), we write

$$a = 2^3 \cdot 3^2 \cdot 5^0 \cdot 7^1 \cdot 17^0 \quad \text{and} \quad b = 2^1 \cdot 3^0 \cdot 5^1 \cdot 7^1 \cdot 17^1.$$

In the notation of the theorem, the a-exponents are $\{a_1, a_2, a_3, a_4, a_5\} = \{3, 2, 0, 1, 0\}$ and the b-exponents are $\{b_1, b_2, b_3, b_4, b_5\} = \{1, 0, 1, 1, 1\}$.

Then c_1 = minimum of a_1 and b_1 = min$\{3, 1\} = 1$.

$c_2 = min\{2, 0\} = 0$

$c_3 = min\{0, 1\} = 0$

$c_4 = min\{1, 1\} = 1$

$c_5 = min\{0, 1\} = 0$

Then $(a, b) = 2^1 \cdot 3^0 \cdot 5^0 \cdot 7^1 \cdot 17^0 = 2 \cdot 7 = 14$.

Figure 4.1   GCD of factored integers.

Proof: See Exercise 4.3.

EXERCISE 4.3. Prove the corollary with the notation of Theorem 4.2. Then compare it with Exercise 3.1.

EXERCISE 4.4. Let $a = 23^5 \cdot 29^8 \cdot 31 \cdot 37^2$ and $b = 29^7 \cdot 37 \cdot 41^2$. Find (a, b).

EXERCISE 4.5. Let $a = 2^8 \cdot 3^5 \cdot 7^9$. Find the smallest positive b for which $(a, b) = 2^6 \cdot 3$.

EXERCISE 4.6. Let $a = 2^4 \cdot 3 \cdot 5$ and $b = 7 \cdot 11 \cdot 13 \cdot 17$. Show that $(a, b) = 1$.

Corollary 4.2 *Two positive integers a and b are relatively prime if and only if their factorizations have no primes in common.*

Proof: See Exercise 4.7.

EXERCISE 4.7. Prove Corollary 4.2 as suggested by Exercise 4.6.

Definition 4.1 *Let a and b be positive integers. Their **least common multiple**, LCM, [a, b], is a multiple of a and a multiple of b and it divides every common multiple of a and b. For convenience we often let [a, b] stand for the positive least common multiple, analogous to (a, b) for the positive greatest common divisor.*

EXAMPLE $[4, 6] = 12$

$$[10, 45] = 90$$

$$[2, 3] = 6$$

$$[2^3 \cdot 3^2 \cdot 7, 2 \cdot 5 \cdot 7 \cdot 17] = 2^3 \cdot 3^2 \cdot 5 \cdot 7 \cdot 17 \text{ (See Theorem 4.3.)}$$

The "lowest common denominator" for two fractions is the LCM of the two denominators.

EXERCISE 4.8. To add 5/12 and 7/18, we need a common denominator for the two fractions. Show that $12 \cdot 18$ can be used as a common denominator, add the fractions, and reduce the sum to lowest terms. Then show that $2^2 \cdot 3^2$ can be used as a common denominator. Show that $2^2 \cdot 3^2$ is $[12, 18]$.

Theorem 4.3   *Let the positive integers a and b be expressed in prime-power factors with the notation of Theorem 4.2. Then $[a, b] = p_1^{m_1} p_2^{m_2} \cdots p_r^{m_r}$, where each $m_i = max\{a_i, b_i\}$.*

Proof: See Exercise 4.9.

EXERCISE 4.9.   Prove Theorem 4.3, referring to the proof of Theorem 4.2.

EXERCISE 4.10.   Referring to Exercise 4.1, find $[726, 1431]$.

EXERCISE 4.11.   Find $(8349, 22)$ and $[8349, 22]$.

Theorem 4.4   *For positive integers a and b, $(a, b)[a, b] = ab$.*

Proof: See Exercise 4.12.

EXERCISE 4.12.   Use Theorems 4.2 and 4.3 to prove Theorem 4.4.

EXERCISE 4.13.   Verify Theorem 4.4 for $a = 726$, $b = 1431$, using Exercises 4.1 and 4.10.

EXERCISE 4.14.   Verify Theorem 4.4 for $a = 8349$, $b = 22$, using Exercise 4.11.

Points to Remember

1.   Each non-zero integer has a factorization into prime powers. This factorization is unique, if order of factors and unit factors are prescribed by convention. On the basis of the laws of arithmetic, unique factorization can be *proved* to be a property of the integers.

2.   The GCD and the LCM of two positive integers can be found from their factorizations by using the minimum and the maximum exponent for each prime power, respectively.

CHAPTER 5

Congruences

5.1 CLOCK ARITHMETIC, A KIND OF EQUIVALENCE

Definition 5.1 *Two integers a and b are* **congruent modulo a positive integer m** *if* $m \mid (a - b)$. *In symbols we write* $a \equiv b \pmod{m}$.

EXAMPLE. 3 and 24 are congruent modulo 7, because 7 divides the difference $3 - 24 = -21$. Also, $3 \equiv 24 \pmod{3}$, because $3 \mid -21$.

175 and 2,097 are congruent modulo 2, because 2 divides the difference $175 - 2,097 = -1,922$.

3 and -5 are congruent modulo 2, because 2 divides the difference $3 - (-5) = 8$. Also, $3 \equiv -5 \pmod{4}$ and $3 \equiv -5 \pmod{8}$.

4 and -3 are not congruent modulo 2, because 2 does not divide $4 - (-3) = 7$.

From Definition 2.1 of division, $m \mid (a - b)$ means that there is a quotient integer q for which $a - b = qm$, or, transposing a term,

$$a = b + qm, \text{ for some integral } q.$$

Then $a \equiv b \pmod{m}$ if and only if a appears in the arithmetic progression

$$..., b - 2m, b - m, b, b + m, b + 2m, b + 3m, ...$$

Congruence arithmetic may seem new, but the notions involved are really quite familiar and you will find you have been applying them for some time without having any mathematical language to describe what you were doing. A familiar instance of congruence is "clock arithmetic". Any two hours differing by a multiple of 12 hours are given the same numeral, so we say that the hours are counted *modulo 12*. In fact, any periodicity gives rise to a congruence relation. Two calendar dates separated by a multiple of 7

| 2 a.m. Monday | 2 p.m. Monday | 2 p.m. Wednesday |

Figure 5.1   Different hours represented alike.

fall on the same day of the week. Two angles differing by a multiple of 360° have common initial and terminal sides.

EXERCISE 5.1.   Write two positive and two negative integers that are congruent to 5 (mod 9).

EXERCISE 5.2.   Write two moduli m for which 5 and 11 are congruent and two moduli for which they are incongruent.

EXERCISE 5.3.   Take as modulus $m = 1$. Then which integers are congruent to each other?

EXERCISE 5.4.   Suppose July 4 falls on Friday. Then give the dates of the Fridays in July and August of the same year.

EXERCISE 5.5.   If the full moon appears next Tuesday, on what day of the week will the moon again be full 28 days later?

EXERCISE 5.6.   Are there any positive integers not equal to 2 and less than 12 that are congruent to 2 (mod 12)?

EXERCISE 5.7.   Are all even numbers congruent to each other modulo 4?

EXERCISE 5.8.   Write two different integers that are congruent (mod 18). Show that they are also congruent (mod 2), (mod 3), (mod 6), and (mod 9). Are there any other moduli for which they are congruent? Find a pair of different integers that are congruent with respect to another modulus besides 18 and its divisors.

EXERCISE 5.9.   Show that the angles 90°, −270°, and 450° are congruent in plane trigonometry. What is the modulus in this case?

EXERCISE 5.10.   An integer is congruent to 6 (mod 10). Is it even?

The symbol ≡ for congruence suggests the familiar = for equality, and indeed both are examples of what we call "equivalence relations".

Definition 5.2 *An* **equivalence relation** *tells, for every two objects covered, whether they are equivalent or not, and also is:*

 i. **reflexive**   *(Each object is equivalent to itself.)*

 ii. **symmetric**   *(If a is equivalent to b, then b is equivalent to a.)*

 iii. **transitive**   *(If a is equivalent to b and b to c, then a is equivalent to c.)*

Some examples of equivalence relations are provided by geometric similarity, geometric congruence, "having the same number as" (for sets), "attending the same class as" (for students). Another useful equivalence is the kind of equality defined for rational numbers (fractions) a/b and c/d; that is, $a/b = c/d$ means $ad = bc$. [For instance, $2/3 = 4/6$.] Some relations that fail in one respect or another to be equivalence relations help clarify the requirements:

"Less than," $<$, among the integers, is

 i. not reflexive [$2 \not< 2$]

 ii. not symmetric [$2 < 3$, but $3 \not< 2$]

 iii. transitive [For all integers a, b, c, $a < b$ and $b < c$ implies $a < c$.]

"less than or equal to," $\leqslant$, among the integers, is

 i. reflexive [$a \leqslant a$ for each integer]

 ii. not symmetric [$2 \leqslant 3$ but $3 \not\leqslant 2$]

 iii. transitive

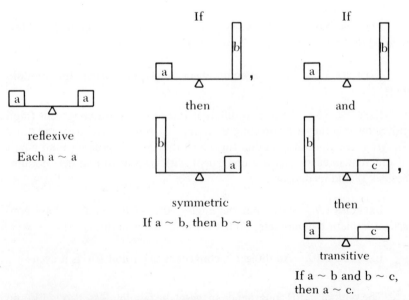

Figure 5.2 The balancing of weights on a beam forms a familiar example of an equivalence relation.

"divides," |, among the positive integers, is

    i. reflexive

    ii. not symmetric

    iii. transitive

Theorem 5.1 *Congruence of integers modulo m is an equivalence relation.*

Proof: Since m is a positive integer (Definition 5.1), it is not zero; thus it is covered by Definition 2.1. Either $m|(a-b)$, in which case $a \equiv b \pmod{m}$, or $m \nmid (a-b)$ and $a \not\equiv b \pmod{m}$. Then congruence has the necessary determining property of an equivalence relation; it does provide a way to tell for every two integers whether they are congruent or not. Also, the relation is

    i. reflexive $[m|(a-a)$ which equals 0, with quotient 0, for each integer $a]$

    i. reflexive $[m|(a-a) = 0$, with quotient 0, for each integer $a]$

    ii. symmetric $[$If $a \equiv b \pmod{m}$, then $m|(a-b)$, so there is a quotient q for which $qm = a-b$. Then $(-q)m = b-a$, so $m|(b-a)$, and $b \equiv a \pmod{m}.]$

    iii. transitive $[$If $a \equiv b$ and $b \equiv c \pmod{m}$, then $m|(a-b)$ and $m|(b-c)$, but $a-c = (a-b)+(b-c)$, so $m|(a-c)$, and $a \equiv c \pmod{m}.]$ ∎

EXERCISE 5.11. Show that equality as defined for fractions is an equivalence relation.

EXERCISE 5.12. For a statistical survey individuals are taken as "equivalent" if they are of the same age. Is this an equivalence relation? For a different survey individuals are taken as equivalent if they are of the same race. Although we can show these relations to be theoretically reflexive, symmetric, and transitive, they may fail to be equivalence relations with respect to the whole population. Why?

EXERCISE 5.13. Write three positive integers and three negative integers that are congruent to 3 modulo 10.

EXERCISE 5.14. Which integers are congruent to 0 modulo 2? to 1 modulo 2? Which integers are congruent to 0 modulo 3? Which are congruent to 0 modulo m?

EXERCISE 5.15. Notice the restriction that the modulus m be a positive integer. If we tried to use a zero modulus what difficulties would we encounter?

EXERCISE 5.16.   Latin students: The positive integer m in congruence is called the *modulus*. Why do we say "modulo m" for "to the modulus m"?

5.2   EQUIVALENCE CLASSES AND THEIR REPRESENTATIVES

Any equivalence relation, congruence modulo m included, partitions the objects it covers into **equivalence classes**, each separate class made up of objects that are equivalent to each other.

For instance, the equivalence $=$ defined for rational numbers partitions the fractions into classes equivalent

to 1/2, as $1/2$, $2/4$, $-1/-2$, $-2/-4$, $10/20$, $500/1000$, ...

to 3/4, as $3/4$, $-3/-4$, $6/8$, $-6/-8$, $9/12$, $-9/-12$, ...

and so on, for each fraction.

We often take the fraction in its lowest terms as representative of the whole equivalence class, and, although the answer to a problem may turn out to be some different member of the equivalence class, we may insist on giving instead that particular representative.

For instance,

$$5/12 + 2/15 = (25+8)/60 = 33/60 = 11/20.$$

For the integers modulo 4 we have four different equivalence classes,

$$\{0, 4, -4, 8, -8, 12, -12, 16, -16, ...\}$$
$$\{1, 5, -3, 9, -7, 13, -11, 17, -15, ...\}$$
$$\{2, 6, -2, 10, -6, 14, -10, 18, -14, ...\}$$
$$\{3, 7, -1, 11, -5, 15, -9, 19, -13, ...\}.$$

Every integer belongs to exactly one of these four residue classes, depending on its remainder, or "residue" when it is divided by 4. No integer belongs to more than one class.

More generally, modulo a positive integer m, we have m different residue classes

$$\{0, m, -m, 2m, -2m, 3m, -3m, ...\}$$
$$\{1, 1+m, 1-m, 1+2m, 1-2m, 1+3m, 1-3m, ...\}$$
$$\{2, 2+m, 2-m, 2+2m, 2-2m, 2+3m, 2-3m, ...\}$$
$$...$$
$$\{m-1, -1+2m, -1, -1+3m, -1-m, -1+4m, -1-2m, ...\}.$$

We can select representatives for the different classes. One useful **complete system of representatives** is the **least non-negative residues** $0, 1, 2, \ldots, m-1$. Modulo 4 the least non-negative residues are 0, 1, 2, and 3. Sometimes with large moduli we use some negative representatives to simplify arithmetic; for instance, modulo 19, we might prefer the complete system $-9, -8, -7, -6, -5, -4, -3, -2, -1, 0, 1, 2, 3, 4, 5, 6, 7, 8, 9$ to the complete system of non-negative representatives 0, 1, 2, 3, 4, 5, 6, 7, 8, 9, 10, 11, 12, 13, 14, 15, 16, 17, 18. As we shall see, members of the same equivalence class are interchangeable in congruence arithmetic, just as 1/2 and 2/4 are interchangeable in the arithmetic of fractions.

EXERCISE 5.17.   Arrange 1, 3, 7, -1, -3, -7, 15, -9, 5, 4 in equivalence classes modulo 8.

EXERCISE 5.18.   Show why 0 and m do not both appear in a complete set of representatives modulo m.

EXERCISE 5.19.   Prove that there is no overlap among residue classes; that is, that no member can fall in two different classes. [What would happen if an integer x did fall in two different residue classes? More generally, there can be no overlap among equivalence classes, no matter what equivalence is used. The proof of this fact follows the same idea as that for congruences.]

EXERCISE 5.20.   Write the least non-negative residues modulo 5. Write the least non-negative residues modulo 8.

EXERCISE 5.21.   A registrar sorts some student records into categories A–E, F–J, K–O, P–T, U–Z, depending on initial of surname. Assuming each surname begins with one of the English alphabet letters, show that this partitioning gives rise to an equivalence "is in the same category as." In general, show that any partitioning of a set into distinct categories [we do not allow categories like A–E and D–J, which would overlap] gives rise to an equivalence. In this chapter we first defined the equivalence "congruence," and then showed how congruence partitions the integers into residue classes. We could, however, have begun by defining the classes and then could have shown that the partitioning induced an equivalence.

Next we want to establish the fact that different representatives of residue classes are interchangeable in congruence arithmetic. For instance, in arithmetic modulo 4 we have

$$
\begin{array}{ccc}
\begin{array}{r} 2 \\ +\ 3 \\ \hline 5 \equiv 1 \end{array} & \text{and} & \begin{array}{r} 2 \\ \times\ 3 \\ \hline 6 \equiv 2. \end{array}
\end{array}
$$

Will we get congruent answers if we combine other representatives congruent respectively to 2 and to 3?

2	−2	−10	14	−10	102
+ 3	+ 7	+ (−9)	+ 15	+ 3	+ 103
5 ≡ 1	5 ≡ 1	1	29 ≡ 1	−7 ≡ 1	205 ≡ 1

2	−2	10	14	−10	102
× 3	× 7	× (−9)	× 15	× 3	× 103
6 ≡ 2	−14 ≡ 2	−90 ≡ 2	210 ≡ 2	−30≡2	10506 ≡ 2

This spot check parallels our experience with clock arithmetic, anticipating the conclusion of interchangeability, and although it does not *prove* interchangeability, as we wish to do, it does indicate how to construct the proof.

Let $2+j4$ stand for any member of the residue class that contains 2. As j takes on all integral values, $2+j4$ traces out the whole residue class. Let $3+k4$ stand for any member of the residue class that contains 3. Then for any two representatives we have

$$\begin{array}{c} 2 + j4 \\ + 3 + k4 \\ \hline 5 + (j+k)4 \\ \equiv 1 \pmod 4 \end{array} \quad \text{and} \quad \begin{array}{c} 2 + j4 \\ \times (3 + k4) \\ \hline 6 + j12 \\ k8 + jk16 \\ \hline 6 + (j3+k2+jk4)4 \\ \equiv 2 \pmod 4 \end{array}$$

Now we generalize from 2 and 3 taken modulo 4 to representatives r and s taken modulo m.

Theorem 5.2 *Representatives of the same equivalence class are interchangeable in congruence arithmetic.*

Proof:

$$\begin{array}{c} r + jm \\ + (s+km) \\ \hline r + s + (j+k)m \\ \equiv r + s \pmod m \end{array} \quad \begin{array}{c} r + jm \\ \times (s+km) \\ \hline rs + jsm \\ krm + jkmm \\ \hline rs + (js+kr+jkm)m \\ \equiv rs \pmod m. \blacksquare \end{array}$$

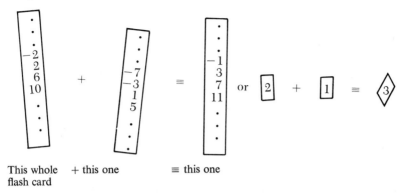

This whole   + this one   ≡ this one
flash card

Figure 5.3   Two Interpretations of Congruence Arithmetic Modulo 4

Since the members of a residue class are interchangeable in addition and multiplication, we can think of combining whole residue classes by combining their respective representatives. Alternatively, we can think of some complete set of residues, such as $0, 1, 2, ..., m-1$, as being all the integers there are modulo m, reducing answers modulo m to this set. It is the latter alternative we use in clocks, for we say there is no 25 o'clock, using 1 o'clock always, even for "15 hours past 10 o'clock."

EXERCISE 5.22.   Complete these addition and multiplication tables modulo 5, using only least non-negative residues.

+	0	1	2	3	4
0	0	1		3	
1	1	2		4	0
2	2			1	
3	3			2	
4			1		

×	0	1	2	3	4
0	0	0	0	0	0
1	0	1	2	3	4
2	0			1	3
3		3		4	
4	0				1

EXERCISE 5.23.   Find the following sums and products; then reduce the numbers in the problems modulo 9, combine them as indicated, and compare with the first answers modulo 9.

$$17,860 + 87,122 =$$

$$4714 + 3913 + 476 + 1324 =$$

$$\$3.17 + \$21.74 + \$18.98 =$$

$$434 \times 28 =$$

$$\$217 \times 19 =$$

EXERCISE 5.24. An experiment is initiated at 3.32 p.m. What time is it 75 hours 48 minutes later?

EXERCISE 5.25. South deals bridge cards one at a time to four players in the order: West, North, East, South, West, and so on. Which player receives the 5th card? the 18th? the 23rd? the 29th? the 33rd? the 44th? the 52nd?

EXERCISE 5.26. A pharmaceutical company deciding on the concentration of a new drug to be taken over a period of several days and nights might weaken the concentration to require administration every 4 hours in preference to every 5 hours, although it would require more frequent doses. Why?

EXERCISE 5.27. Games sometimes begin with "counting out" rhymes for determining who is "it" or who has first turn, and probably many of us got our earliest practice on congruences in this way. Analyze the following two in terms of congruences (or use others you know), showing how the choice will fall among 2 players, 3 players, and m players.

> Did you eever, iver, over
> In your leef, life, loaf
> See the deevil, divil, dovol
> Kiss his weef, wife, woaf?
> No, I neever, niver, nover
> In my leaf, life, loaf
> Saw the deevil, divil, dovol
> Kiss his weef, wife, woaf.

> One potato, two potato, three potato, four:
> Five potato, six potato, seven potato, or.

(The players stand in a circle with their fists upraised. Each successive beat of the verse "counts" the corresponding one of the $2m$ fists.)

EXERCISE 5.28. There are fruitful generalizations of the congruence idea. For instance, consider all polynomials in x with integral coefficients, such as $3x^5 + 4x^3 + 2x - 7$ or $-2x^3 + x + 3$. These polynomials have an arithmetic of their own; we add the coefficients of like powers of x, and we multiply term by term. For example,

$$(3x^5 + 4x^3 + 2x - 7) + (-2x^3 + x + 3)$$

$$= 3x^5 + (4-2)x^3 + (2+1)x + (-7+3)$$

$$= 3x^5 + 2x^3 + 3x - 4.$$

and

$$3x^5 + 4x^3 + 2x - 7$$
$$\times (-2x^3 + x + 3)$$

$$-6x^8 - 8x^6 - 4x^4 + 14x^3$$
$$3x^6 + 4x^4 \qquad\quad + 2x^2 - 7x$$
$$9x^5 \qquad + 12x^3 \qquad\quad + 6x - 21$$

$$-6x^8 - 5x^6 + 9x^5 + 26x^3 + 2x^2 - x - 21$$

Now take these polynomials with respect to two moduli, the integral modulus 2, and a polynomial modulus $x^2 + x + 1$. From the modulus 2, we see that we can always choose as representative of a residue class a polynomial having coefficients 0 and 1. For example, $3x^5 + 4x^3 + 2x - 7 \equiv (3 - 2)x^5 + (4 - 2 \cdot 2)x^3 + (2 - 2)x - 7 + 2 \cdot 4 \equiv x^5 + 1 \pmod 2$, and $-2x^3 + x + 3 \equiv x + 1 \pmod 2$. From the polynomial modulus $x^2 + x + 1$, we see that our representatives need have no terms in x^2, x^3, x^4, or higher powers of x, because we can subtract multiples of $x^2 + x + 1 \equiv 0$ to reduce their degrees. For example, $x^5 + 1 \equiv x^5 - x^3(x^2 + x + 1) + 1 \equiv -x^4 - x^3 + 1 \equiv -x^4 + x^2(x^2 + x + 1) - x^3 + 1 \equiv x^2 + 1 \equiv x^2 - (x^2 + x + 1) + 1 \equiv -x \pmod{x^2 + x + 1} \equiv x \pmod{\text{modulis } 2, x^2 + x + 1}$. A complete system of residues, then, has only four polynomials,

$$0, \quad 1, \quad x, \quad x+1.$$

Complete the addition and multiplication tables for these four, reducing answers modulis 2 and $x^2 + x + 1$.

+	0	1	x	$x+1$
0	0	1	x	
1	1			x
x				1
$x+1$		1		

$\times$	0	1	x	$x+1$
0	0	0		0
1				$x+1$
x	0			1
$x+1$				x

EXERCISE 5.29. Latin students: In Exercise 5.28 note the use of "moduli" as a plural. Can you identify the origin of the form "modulis" for "to more than one modulus"?

5.3 DIVISIBILITY OF LARGE NUMBERS

Before computers took over accounting chores, human adders were taught to check sums by "casting out nines." They summed the digits in each row modulo 9, then checked to see that the sums modulo 9 added up, as illustrated:

$$\$14.32 \qquad 1 + 4 + 3 + 2 \equiv 1 \,(\mathrm{mod}\,9)$$
$$1.48 \qquad 1 + 4 + 8 \equiv 4$$
$$76.25 \qquad 7 + 6 + 2 + 5 \equiv 2$$
$$75.65 \qquad 7 + 5 + 6 + 5 \equiv 5$$
$$+ \;\; 4.13 \qquad 4 + 1 + 3 \equiv 8$$

$$\overline{\$171.83} \qquad \overline{1 + 7 + 1 + 8 + 3 \equiv 2 \,(\mathrm{mod}\,9)}$$

With congruence notation and a knowledge of our decimal system, we can show why "casting out nines" works. In our decimal notation each digit position has the value of a different power of 10; for instance,

$14.32 = 1$ ten-dollar bill $+ 4$ one-dollar bills $+ 3$ dimes $+ 2$ pennies,

or

$$1432 = 1 \times 10^3 + 4 \times 10^2 + 3 \times 10^1 + 2 \times 10^0.$$

Now since $10 \equiv 1 \,(\mathrm{mod}\,9)$, we have $10^d \equiv 1 \,(\mathrm{mod}\,9)$, for each power of 10, so that

$$1432 \equiv 1 \cdot 1^3 + 4 \cdot 1^2 + 3 \cdot 1 + 2 \equiv 1 + 4 + 3 + 2 \,(\mathrm{mod}\,9).$$

Casting out nines is based on congruence properties modulo 9.

This same idea can be used to test a large number for divisibility by 9: Is 5438777 divisible by 9?

$$5438777 \equiv 5 + 4 + 3 + 8 + 7 + 7 + 7$$

$$\equiv 5 \not\equiv 0 \,(\mathrm{mod}\,9),$$

so

$$9 \nmid 5438777.$$

This establishes the first of these divisibility criteria:

1. An integer is divisible by 9 if and only if the sum of its digits is divisible by 9.

2. An integer is divisible by 10 if and only if its units digit is 0. [$10^d \equiv 0 \,(\mathrm{mod}\,10)$ if $d > 0$]

3. An integer is divisible by 2 if and only if its units digit is 0, 2, 4, 6, or 8. [$10^d \equiv 0 \,(\mathrm{mod}\,2)$ if $d > 0$]

4. An integer is divisible by 5 if and only if its units digit is 0 or 5.

5. An integer is divisible by 3 if and only if the sum of its digits is divisible by 3. $[10^d \equiv 1 \pmod 3]$

6. An integer is divisible by 4 if and only if the integer formed of its tens and units digits is congruent to $0 \pmod 4$. $[10^d \equiv 0 \pmod 4$ for $d > 1]$

Some of the following exercises will help you to construct other divisibility criteria.

EXERCISE 5.30.   Test each of the following for divisibility by 2, 3, 4, 5, 9, and 10.

12345	738	23456
37	70006	1002
4275	180	994

EXERCISE 5.31.   Write out 4268 in powers of 10. Now prove that it is divisible by 11 by finding $10 \equiv ? \pmod{11}$, so $10^d \equiv ? \pmod{11}$. Write out a general rule for testing for divisibility by 11.

EXERCISE 5.32.   Use the divisibility criteria for 2 and for 4 to develop a criterion for 8. Notice in this connection that $8 \nmid 124$ but $8 \mid 224$. Use your criterion to check each multiple of 4 in Exercise 5.30 for divisibility by 8.

We can write integers to base 8. For example, the integer 7126 can be written as

$$7126 = 1 \times 8^4 + 5 \times 8^3 + 7 \times 8^2 + 2 \times 8^1 + 6 \times 8^0,$$

or to base 8, $7126_8 = 15726$. The easiest way to get the coefficients of powers of 8 is to divide the integer in stages by higher and higher powers of 8.

$$8 | \underline{7126}$$
$$8 | \underline{890}/6 \times 8^0$$
$$8 | \underline{111}/2 \times 8^1$$
$$8 | \underline{13}/7 \times 8^2$$
$$1/5 \times 8^3$$

In base 8 only the eight digits $0, 1, 2, 3, 4, 5, 6, 7$ appear. How is 8 written to base 8? 9 to base 8?

For bases larger than 10 we need extra digits. For instance, in the base 12 system we need a digit such as T for 10 and a digit such as E for 11. Write the integer 4168 to base 8 and to base 12.

EXERCISE 5.33. How can we test an integer written to base 8 for divisibility by 8? by 4? by 2? by 16? by 7? [Recall casting out 9's to base 10.]

EXERCISE 5.34. How can we test an integer written to base 12 for divisibility by 12? by 3? by 4? by 11?

EXERCISE 5.35. How can we test an integer written to base b for divisibility by b? by $b-1$? by a divisor d of b?

Points to Remember

1. Congruence is an equivalence relation, as are similarity among triangles and fraction-equality among rational numbers.

2. Congruence is especially well adapted to treating periodicity. As n increases through positive integers, the remainder left when we divide n by a modulus $m > 1$ is periodic. Congruence modulo m makes equivalent all n's that leave the same remainder when divided by m.

SUGGESTED PROJECTS

1. The various tests for divisibility, by 10, 100, 1000, 5, 2, 4, 8, and so on, can be collected to form a report or a talk. If the audience has worked with change of base (also see Project at end of Chapter 1), then it is appropriate to develop divisibility tests for numbers written to other bases, as in Exercises 5.32 through 5.35. This makes a good link with the tests for divisibility in decimal notation, because it points up why our usual tests concern divisibility by 10, its powers, and its divisors.

2. Congruence arithmetic can be enjoyable as a relief from the larger numbers of ordinary arithmetic. A modulus of 4 provides a good introduction. For an elementary presentation, demonstrate how to add and multiply modulo 4, "throwing away" multiples of 4. As an example you might use dollars and quarters taken in for theater tickets. The amounts taken in, reduced modulo 4, tell how many coins there are after all possible bunches of four quarters have been converted to dollars. You can use this example to make $2 + 2 \equiv 0$ and $2 \cdot 6 \equiv 0$ look plausible and even useful. Then contrast ordinary arithmetic with congruence arithmetic on the basis of zero products, since modulo 4 the product $2 \cdot 6$ is zero, but neither factor is zero.

CHAPTER 6

Solving Congruences

We can write all the integers congruent to a given integer b modulo m by writing $b + qm$, where q is any integer. This problem might be stated as a congruence to be solved for the values of x that make it hold:

$$x \equiv b \pmod{m}.$$

This is a "conditional" congruence analogous to a "conditional" equation like

$$x^2 - x - 56 = 0.$$

The equation does not represent a number statement supposed to hold for all numbers, or all integers, but may hold on "condition" that x take on certain "solution" values. (Is $x^2 > 0$ a number statement that holds for all integers? for all non-zero integers? For what integers is $x^2 < 0$ a valid number statement?)

We say that we "solve" the equation $x^2 - x - 56 = 0$ when we find all the numbers, -7 and $+8$, that "satisfy" the equation; each of these solution numbers has the property that if we square it, then subtract the number, and then subtract 56, we get zero. This particular equation can be solved by factoring the left member

$$(x + 7)(x - 8) = 0$$

or by using the quadratic formula

$$x_1 = \frac{-b + \sqrt{b^2 - 4ac}}{2a}, \qquad x_2 = \frac{-b - \sqrt{b^2 - 4ac}}{2a}$$

$$= \frac{+1 + \sqrt{1 + 4 \cdot 56}}{2} \qquad = \frac{1 - 15}{2}$$

$$= \frac{1 + 15}{2} = 8 \qquad = -7$$

However, equations of higher degree may be quite difficult to solve; in fact, some may require more than a finite number of steps, as there is no formula solution in general for equations of degree higher than 4.

We have a much simpler problem if we are asked to solve a *congruence*, such as

$$x^2 - x - 56 \equiv 0 \text{ modulo } 11.$$

As in solving an equation, we "solve" a congruence by finding all the integers that "satisfy" the congruence. In the example here we want integers such that the square minus the integer minus 56 is an integer congruent to 0 (mod 11), that is, divisible by 11.

From Theorem 5.2 congruence arithmetic gives the same results no matter what representative we select from an equivalence class. For this reason we need try only 11 integers, a representative of each residue class modulo 11. Thus we can find all solutions to the congruence by trial and error.

$(0)^2 - (0) - 56 \equiv -1 \not\equiv 0 \pmod{11}$

$(1)^2 - (1) - 56 \equiv -1$ $\qquad$ $(6)^2 - (6) - 56 \equiv -4$

$(2)^2 - (2) - 56 \equiv 1$ $\qquad$ $(7)^2 - (7) - 56 \equiv -3$

$(3)^2 - (3) - 56 \equiv 5$ $\qquad$ $(8)^2 - (8) - 56 \equiv 9 - 8 - 1 \equiv 0$

$(4)^2 - (4) - 56 \equiv 5 - 4 - 1 \equiv 0$ $\qquad$ $(9)^2 - (9) - 56 \equiv -6$

$(5)^2 - (5) - 56 \equiv -3$ $\qquad$ $(10)^2 - (10) - 56 \equiv 1$

We find by experiment the two solutions $x_1 \equiv 4$ and $x_2 \equiv 8$. The solutions are written in congruence form, too, because any other representatives of their respective residue classes are solutions. The related equation has a solution -7, which is in the same residue class as 4 modulo 11.

EXERCISE 6.1. Solve the congruence $x^3 - x^2 + 7x + 1 \equiv 0 \pmod{8}$ by trial and error. Although the degree of this congruence is only 3, it has four solution classes.

6.1 LINEAR CONGRUENCES

A congruence of degree 1 is called a **linear** congruence. Its general form is

$$ax \equiv b \pmod{m}, \qquad a \not\equiv 0 \pmod{m}.$$

It is instructive to solve some linear congruences by trial and error before we develop more systematic ways to solve them.

EXAMPLE 1.   Solve $3x \equiv 1 \pmod{14}$. We find $3(1) \equiv 3$, $3(2) \equiv 6$, $3(3) \equiv 9$, $3(4) \equiv 12$, **$3(5) \equiv 1$**, $3(6) \equiv 4$, $3(7) \equiv 7$, $3(8) \equiv 10$, $3(9) \equiv 13$, $3(10) \equiv 2$, $3(11) \equiv 5$, $3(12) \equiv 8$, $3(13) \equiv 11$. There is one solution class, the equivalence class that contains 5.

EXAMPLE 2.   Solve $4x \equiv 3 \pmod{14}$. We note first that $4(1) \equiv 4$, $4(2) \equiv 8$, $4(3) \equiv 12$, and so on. We notice that $4x$ is even for each x, so that $4x - q14$ is even for any q. Then there can be no x for which $4x$ is congruent to 3, an odd number. This linear congruence has no solution classes.

EXAMPLE 3.   Solve $4x \equiv 6 \pmod{14}$. By trial and error we find two solution classes, the one containing 5 and the one containing 12.

We can develop a general theory from the analysis suggested by Example 2.

EXERCISE 6.2.   Try to state and prove a conjecture about linear congruences $ax \equiv b \pmod{m}$ in case $(a, m) \nmid b$. Then check your result with the following theorem.

Theorem 6.1   *Let $(a, m) \nmid b$ in the linear congruence*

$$ax \equiv b \pmod{m}.$$

Then the congruence has no solutions.

Proof: By Definition 5.1, if s is a solution, so that $a \cdot s \equiv b \pmod{m}$, this means that

$$m \mid (as - b),$$

so that

$$as - b = mq$$

for some q. Then $b = as - mq$, which is divisible by every common divisor of a and m, including (a, m). But by hypothesis $(a, m) \nmid b$. Then there is no solution s. ∎

Corollary 6.1   *In congruence arithmetic modulo m an integer a has no multiplicative inverse a^{-1} such that $a \cdot a^{-1} \equiv 1$ unless a is relatively prime to m.*

Proof: See Exercise 6.3.

EXERCISE 6.3.   Show that $ax \equiv 1 \pmod{m}$ has no solution if a is not relatively prime to m.

In both Example 1 and Example 3 we note that $(a, m)|b$, but in the latter example $(a, m) > 1$, and the congruence has more than one solution.

Theorem 6.2   *Let $(a, m) = d$ in the linear congruence*

$$ax \equiv b \pmod{m}$$

and let $d|b$. Let $m = m'd$. Then the congruence has d distinct solutions (mod m), $x_1, x_2, \ldots, x_d$, with $(i-1)m' \leqslant x_i = x_1 + (i-1)m' < im', i = 1, 2, \ldots, d$.

Proof: Suppose s is a solution to the congruence

$$ax \equiv b \pmod{m}; \tag{1}$$

that is,

$$as \equiv b \pmod{m}.$$

This means that $m|(as-b)$, or, since d divides a, b, and m,

$$m'd|(a'ds-b'd),$$

where $a = a'd$ and $b = b'd$. But this happens if and only if

$$m'|(a's-b'),$$

that is, if and only if s is a solution for the congruence

$$a'x \equiv b' \pmod{m'}. \tag{2}$$

Now $(a', m') = 1$, by Corollary 4.1. In case $(a, m) = d = 1$, congruence (2) coincides with congruence (1).

Now we solve congruence (2). Since $(a', m') = 1$, we have, by Theorem 2.3, integers r and s for which

$$1 = ra' + sm'.$$

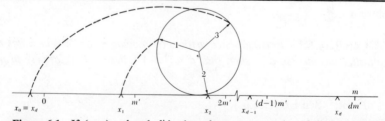

Figure 6.1   If $(a, m) = d$ and $d|b$, then the congruence has d distinct solutions (mod m), one in each interval of length $m' = m/d$.

Since $m' \equiv 0 \pmod{m'}$, we can write from the equation a congruence

$$1 \equiv ra' \pmod{m'}.$$

Then rb' supplies a solution to the congruence (2), for

$$a'(rb') = (a'r)b' \equiv (1)b' \equiv b' \pmod{m'}.$$

We have found one solution for (2). Are there any others $\pmod{m'}$? Suppose s_1 and s_2 are two solutions. Then $a's_1 \equiv b'$ and also $a's_2 \equiv b' \pmod{m'}$, so, subtracting,

$$a'(s_1 - s_2) \equiv 0 \pmod{m'}.$$

Then $ra'(s_1 - s_2) \equiv (1)(s_1 - s_2) \equiv s_1 - s_2 \equiv 0 \pmod{m'}$. Then there is exactly one solution x_1 in the range $0 \leqslant x_1 < m'$.

Each $x_i = x_1 + (i-1)m'$, $i = 1, 2, ..., d$, is a solution for (1), since

$$ax_i = a[x_1 + (i-1)m'] = ax_1 + a(i-1)m'$$
$$= da'x_1 + a'(i-1)dm'$$
$$= db' + a'(i-1)m$$
$$\equiv b \pmod{m}. \quad \blacksquare$$

See Figure 6.2 for a numerical example.

Corollary 6.2 *In congruence arithmetic modulo m, a has a multiplicative inverse if a is relatively prime to m.*

Proof: See Exercise 6.4.

EXERCISE 6.4. Use Theorem 6.2 with $b = 1$ to prove the corollary.

Solve (1) $6x \equiv 9 \pmod{15}$. Since $(6, 15) = 3 \,|\, 9$, we divide the congruence by 3 to get (2) $2x \equiv 3 \pmod 5$.

Here $(2, 5) = 1 = (-2)2 + (1)5$, so that $1 \equiv (-2)2 \pmod 5$. Then $x_1 \equiv (-2)3 \equiv 4$ is a solution for congruence (2):

$$2x_1 \equiv 2[(-2)3] \equiv [2(-2)]3 \equiv 1 \cdot 3 \equiv 3 \pmod 5.$$

The solutions $\pmod{15}$ for congruence (1) are the classes having representatives

$$x_1 \equiv 4, \qquad x_2 \equiv 4 + 5 \equiv 9, \qquad x_3 \equiv 4 + 2 \cdot 5 \equiv 14.$$

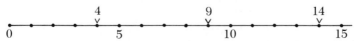

Figure 6.2 Solution of the congruence $6x \equiv 9 \pmod{15}$.

EXERCISE 6.5.   Find all solution classes for $10x \equiv 6 \pmod{18}$. Draw a chart like the one in Figure 6.2.

EXERCISE 6.6.   Find all solution classes for $3x \equiv 2 \pmod 7$.

EXERCISE 6.7.   Find all solution classes for $200x \equiv 10 \pmod{35}$.

EXERCISE 6.8.   A shopper needs 4 cans of pet food each day, but wants to take advantage of a 6-can-for-$1 offer. What numbers of cans can she buy to satisfy these requirements? What is the smallest number of cans that will do?

EXERCISE 6.9.   I bought 3 loaves of bread, paid in dollar bills, and got 89¢ change. How much did each loaf cost?

EXERCISE 6.10.   The spy had just time enough to note that the 50-place dial of the safe was set on 2 when he heard the counterspy approach. He hid and could see the counterspy turn the dial 3 times clockwise, an equal (from the clicks) but unknown number x of places each time. Then the counterspy was interrupted and fled, and the spy noted that the dial was at 7. How far did the counterspy turn the dial each time?

EXERCISE 6.11.   Suppose that -22 and 23 are congruent modulo m. Find all possible values for the modulus m. Find two values for m for which $-22 \not\equiv 23$.

EXERCISE 6.12.   Find which Presidential inauguration years (that is, the year after the Presidential election) end in the digit 7.

6.2   SIMULTANEOUS LINEAR CONGRUENCES

EXERCISE 6.13.   Find which solution classes for $5x \equiv 3 \pmod 7$ are also solutions for $2x \equiv 1 \pmod 5$.

EXERCISE 6.14.   Every fourth week it is my turn to drive in a car pool and every sixth week I am responsible for a club program. This week they coincide. How often will that happen? Would there be any arrangement for starting these two cyclical duties so that they would never fall in the same week?

With the following theorem we can determine the interference pattern of several periodic events. Sometimes we want to know whether we can schedule periodic events deliberately so as to avoid interference, as suggested in Exercise 6.14, and sometimes we want events to coincide or want to know when and how often they will coincide.

Theorem 6.3   *The congruences* $x \equiv x_1 (mod\ m_1)$ *and* $x \equiv x_2 (mod\ m_2)$ *have no simultaneous solution if* (m_1, m_2) *fails to divide* $(x_1 - x_2)$. *If* (m_1, m_2) $= d$ *divides* $(x_1 - x_2)$, *there is exactly one solution modulo* $[m_1, m_2]$.

The n congruences $x \equiv x_i (mod\ m_i)$, $i = 1, 2, ..., n$, *have a simultaneous solution if and only if each pair have a simultaneous solution, in which case there is exactly one solution modulo the least common multiple of all the moduli, as defined in Exercise 6.18. When each pair of moduli are relatively prime, this gives the famous "Chinese Remainder Theorem."*

Notice that any system of simultaneous linear congruences can be reduced to the form called for in Theorem 6.3, for each linear congruence $ax \equiv b(mod\ m)$ has solutions, and hence simultaneous solutions with other congruences, only if (a, m) divides b, in which case it can be replaced by $x \equiv (a')^{-1}b'(mod\ m')$, where primes denote division by (a, m).

The Chinese Remainder Theorem can be used on some ancient puzzles similar to Exercise 6.15 below, and has also been used to find when heavenly bodies in their respective periodic orbits will lie in a given relationship.

EXERCISE 6.15.   The baker tries to arrange x birthday candles in rows. If he uses rows of 2, there is one candle left over. If he uses rows of 3, there are 2 left. If he uses rows of 4 there are 3 left, but he finds that if he uses rows of 5 the candles come out even. How many candles are there?

EXERCISE 6.16.   In years whose numbers are divisible by 4, an adjustment is made to our calendar by adding Leap Year Day, but in years whose numbers are divisible by 100 a finer adjustment is made by leaving it out. Show by Theorem 6.3 that the finer adjustment does occur, and find the most recent such year.

Optional Proof: Since the proof of Theorem 6.3, though elementary, is complicated, we recommend that you skip it on a first reading and go directly to Chapter 7.

Proof for two simultaneous congruences: Let $m_1 = m_1'd$ and $m_2 = m_2'd$. From the definition of congruence expressed in the Division Algorithm form, if $\bar{x}$ is a simultaneous solution for $x \equiv x_1 (mod\ m_1)$ and $x \equiv x_2 (mod\ m_2)$, then $\bar{x} = x_1 + q_1 m_1$ and also $\bar{x} = x_2 + q_2 m_2$, so these two expressions must be equal. Transposing terms, we have

$$x_1 - x_2 = q_2 m_2 - q_1 m_1 = (q_2 m_2' - q_1 m_1')d. \tag{1}$$

From (1), $d | (x_1 - x_2)$ if there is any solution $\bar{x}$. Then suppose $d | (x_1 - x_2)$, which we can write in division form as $x_1 - x_2 = qd$, for some integral quotient q. The Euclidean Algorithm can be used to write d linearly as

$d = rm_1 + sm_2$. Choose $\bar{x} = x_1 - qrm_1$. Then $\bar{x} \equiv x_1 \pmod{m_1}$, so that $\bar{x}$ is a solution for the first congruence. But $x_1 - qrm_1 = (x_2 + qd) - qrm_1 = x_2 + q(rm_1 + sm_2) - qrm_1 = x_2 + qsm_2 \equiv x_2 \pmod{m_2}$. Then $\bar{x}$ is also a solution for the second congruence.

Suppose $\bar{x}$ and $\bar{y}$ are both simultaneous solutions for the two congruences. Then from $\bar{x} \equiv x_1 \pmod{m_1}$ and $\bar{y} \equiv x_1 \pmod{m_1}$, we have $m_1 | (\bar{x} - \bar{y})$. Similarly, since both are solutions for the second congruence, we have $m_2 | (\bar{x} - \bar{y})$. Then the difference $\bar{x} - \bar{y}$ is a common multiple of m_1 and m_2, hence is divisible by their least common multiple $[m_1, m_2]$. Then there can be no more than one solution in each interval of length $[m_1, m_2]$.

Suppose $\bar{x}$ is a common solution. Then there is a common solution in each interval of length $[m_1, m_2]$, because each $\bar{x} + k[m_1, m_2]$, where k is an integer, provides a solution, since $[m_1, m_2] \equiv 0$ for each modulus.

The following exercise sequence shows how to supply a proof that n congruences $x \equiv x_i \pmod{m_i}$, $i = 1, 2, ..., n$ have a simultaneous solution if and only if $(m_i, m_j) | (x_i - x_j)$ for every pair $i < j$.

EXERCISE 6.17.   Suppose that among n congruences, with $n \geqslant 2$, some pair, say the first two, have $(m_1, m_2) \nmid (x_1 - x_2)$. Then prove there can be no simultaneous solution for the n congruences.

EXERCISE 6.18.   Define a least common multiple for more than two integers as a common multiple of all the integers that is divisible by every common multiple (cf. Definition 4.1). Extend Theorem 4.3 to cover this case.

EXERCISE 6.19.   To prove Theorem 6.3 inductively, that is, stepwise, assume that we have proved that $k-1$ of the n given congruences have a simultaneous solution $\bar{x}$ modulo the least common multiple L of the given moduli. That is,

$$x \begin{cases} \equiv \bar{x} \bmod L, \\ \equiv x_1 \pmod{m_1}, x_2 \pmod{m_2}, ..., x_{k-1} \pmod{m_{k-1}}, \end{cases}$$

where $2 \leqslant k-1 < n$.

We want to show that the two congruences

$$x \equiv \bar{x} \pmod{L}$$

$$x \equiv x_k \pmod{m_k}$$

have a simultaneous solution.

From the first part of Theorem 6.3, already established, these two congruences have a simultaneous solution if and only if the greatest common divisor $g = (L, m_k)$ divides $\bar{x} - x_k$. Let g be written in its factored form as a product of prime-powers p^b. Find L and g for the special case $m_1 = 12$, $m_2 = 18$, $m_k = m_3 = 54$. Using the extension of Theorem 4.3 suggested in Exercise 6.18, show that each p^b divides m_k and also divides m_i for some $i < k$, and hence $p^b \mid (m_i, m_k) \mid (\bar{x} - x_k) \equiv x_i - x_k \pmod{m_i}$.

EXERCISE 6.20.   Complete the proof of Theorem 6.3.   ∎

Points to Remember

1.   We can tell whether a linear congruence $ax \equiv b \pmod{m}$ has solutions by whether (a, m) divides b. If $d = (a, m)$ divides b, then there are d solutions modulo m.

2.   An integer a has a multiplicative inverse a^{-1} modulo m if and only if $(a, m) = 1$.

3.   We can tell by reference to Theorem 6.3 whether simultaneous linear congruences have a solution.

SUGGESTED PROJECT

Simple trial-and-error solution of linear congruences requires very little in the way of introduction. Discovery that some congruences cannot have a solution, and why, encourages analysis of each case. For an elementary presentation a numerical experimental approach is indicated.

The oldest methods of solving equations were of this experimental nature. To find what the cost of the carrots was in a word problem, our ancestors said, "Try a cost of 15¢. If the result is too large, try a smaller price. If it is too small, try a larger price." This approach has much to recommend it, as it leads naturally to understanding of approximations, interpolation, and extrapolation. It tends to make people confident that in an actual problem they will be able to find the answer.

Another approach to solving a congruence is through a graph. The congruence

$$ax \equiv b \pmod{m}$$

means

$$ax = b + ym, \quad \text{for some integer } y,$$

or

$$ax - my = b, \quad \text{with } x \text{ and } y \text{ integers.}$$

Draw on a cartesian graph the line having formula

$$5x - 10y = 15.$$

Demonstrate that each solution x' for the congruence

$$5x \equiv 15(\bmod 10)$$

corresponds to a lattice point (x', y') on the line, that is, a point at which the line crosses an intersection of the cartesian grid.

Draw on another graph the line having formula

$$5x - 10y = 2.$$

Note that this line touches none of the lattice points of the cartesian grid, which corresponds graphically to the algebraic fact that

$$5x \equiv 2(\bmod 10)$$

has no solutions, since $(5, 10) \nmid 2$.

CHAPTER 7

Elements Prime to the Modulus

7.1 THE EULER PHI FUNCTION

If we combine the results of Corollaries 6.1 and 6.2, we have:

An integer a has a multiplicative inverse a^{-1} modulo m if and only if a is relatively prime to m.

The property of having an inverse is so important that we single out these integers relatively prime to m for special attention. Our first objective is to count them. How many of the integers that are different modulo m are relatively prime to m?

EXERCISE 7.1. Prove that an integer b is relatively prime to the modulus m if and only if every integer in the equivalence class of b modulo m is relatively prime to m.

Definition 7.1 *The **Euler phi function**, $\phi(n)$, of a positive integer n is the number of positive integers less than or equal to n that are relatively prime to n. [Since $(n, n) = n$, n is relatively prime to n only if $n = 1$.]*

EXERCISE 7.2. Find $\phi(1)$, $\phi(2)$, ..., for the first 13 positive integers. A sample is given in Figure 7.1.

EXERCISE 7.3. Find $\phi(p)$ for five different primes p. Form a conjecture about the value of $\phi(p)$, where p is a prime.

Theorem 7.1 *For a prime p the function $\phi(p) = p - 1$.*

Proof: Let b be an integer with $0 < b < p$. If $(b, p) = d$, then $d|p$ and $d \leqslant b < p$. Then since p is a prime, Theorem 3.2 implies that $d = 1$. Then by Theorem 3.1, b is relatively prime to p. Then the $p - 1$ positive integers less than p are prime to p. ∎

Find $\phi(15)$.

1, prime to 15	$(6, 15) = 3$	$(11, 15) = 1$
2, prime to 15	$(7, 15) = 1$	$(12, 15) = 3$
3, $(3, 15) = 3$	$(8, 15) = 1$	$(13, 15) = 1$
4, prime to 15	$(9, 15) = 3$	$(14, 15) = 1$
5, $(5, 15) = 5$	$(10, 15) = 5$	$(15, 15) = 15$
	$\phi(15) = 8$	

Figure 7.1

EXERCISE 7.4.   Find $\phi(3)$, $\phi(9)$, $\phi(27)$, and $\phi(81)$, and form a conjecture about the value of $\phi(p^w)$, where p^w is a power of a prime p.

Theorem 7.2 *Let p^w be a power of a prime. Then $\phi(p^w) = p^{w-1}(p-1)$.*

Proof: Let b be an integer with $0 < b < p^w$, and let $(b, p^w) = d$. Either $d = 1$, in which case b is prime to p^w, or d is a power of p, since $d|p^w$. Then the positive integers less than p^w that are not prime to p^w are those that are divisible by p, that is, the set

$$\{p, 2p, ..., p^w - p\},$$

whose members are p times the multipliers $\{1, 2, ..., p^{w-1} - 1\}$, hence $p^{w-1} - 1$ in number. From the $p^w - 1$ positive integers less than p^w, we subtract the $p^{w-1} - 1$ that are not prime to p^w, leaving $p^w - p^{w-1} = p^{w-1}(p-1)$ that are prime to p^w. ∎

EXERCISE 7.5.   Prove Theorem 7.2 by induction on the exponent w. What theorem supplies the basis for the induction, the proof that Theorem 7.2 holds for $w = 1$? Next, take as the induction hypothesis: For $w = k$, $\phi(p^k) = p^{k-1}(p-1)$. The least positive residues taken modulo p^{k+1} are the integers $b, b+p^k, b+2p^k, ..., b+(p-1)p^k$, where b ranges over the least positive residues (mod p^k). [Verify for 3^3 and 3^2.] All p integers $b + tp^k$ with $0 \leqslant t < p$ are relatively prime to p^{k+1} if and only if b is relatively prime to p^k. Therefore there are p times as many integers prime to p^{k+1}, or $p[p^{k-1}(p-1)]$ $= p^k(p-1)$. ∎

Now, from Chapter 4 we can always think of a modulus m as a product of prime powers. This fact makes it possible for us to use the Chinese Remainder Theorem 6.3 to evaluate $\phi(m)$.

Theorem 7.3   *Let* $m = p_1^{a_1} p_2^{a_2} \ldots p_k^{a_k}$, *where the primes* p_i *are distinct and the exponents are positive. Then* $\phi(m) = \phi(p_1^{a_1}) \phi(p_2^{a_2}) \ldots \phi(p_k^{a_k})$.

EXAMPLE.

$$\phi(180) = \phi(2^2 \cdot 3^2 \cdot 5) = \phi(2^2) \cdot \phi(3^2) \cdot \phi(5)$$

$$= 2(2-1) \cdot 3(3-1) \cdot (5-1) = 48.$$

Proof: An integer b is prime to m if and only if it is prime to each of its prime-power factors p^a; for, if one of the prime factors p divides b, then $p \mid (b, m)$. Conversely, if b is not relatively prime to m so that $(b, m) > 1$, then (b, m) has a prime divisor, say p, with b not prime to p^a. Then to find an integer b that is prime to m we must find one that is prime to each factor p^a. Another way to say this is that b must be congruent modulo each $p_i^{a_i}$ to an integer b_i that is prime to $p_i^{a_i}$.

We can write this requirement as

$$b \equiv \begin{cases} b_1 \,(\mathrm{mod}\, p_1^{a_1}) \\ b_2 \,(\mathrm{mod}\, p_2^{a_2}) \\ \vdots \\ b_k \,(\mathrm{mod}\, p_k^{a_k}) \end{cases}$$

where b_i is one of the $\phi(p_i^{a_i})$ integers prime to $p_i^{a_i}$, $i = 1, 2, 3, \ldots, k$. Are there simultaneous solutions? Yes, for in the requirement of Theorem 6.3 we have for this case $(p_i^{a_i}, p_j^{a_j}) = 1$ for $i \neq j$. There is exactly one solution modulo the least common multiple m of the prime powers for each choice of the b_i's, and there are $\phi(p_1^{a_1}) \phi(p_2^{a_2}) \ldots \phi(p_k^{a_k})$ such choices. ∎

EXERCISE 7.6.   Verify Theorem 7.3 where possible in Exercise 7.2.

With the factorization theory of Chapter 4 available to us, we have jumped immediately to m in its form as a product of prime powers. We could have used the Chinese Remainder Theorem 6.3 in exactly the same way to show that if m is expressed as a product of relatively prime factors,

$$m = rs, \quad \text{where } (r, s) = 1,$$

then $\phi(m) = \phi(r) \phi(s)$. Such a function is said to be **multiplicative**.

7.2 EXPONENTIATION

In ordinary arithmetic we get little practice in raising numbers to powers, because the results increase so rapidly as the exponent increases that the computations become prohibitive. For instance, even a small number like 5 raised to small exponents gives us

$$5^2 = 25, \quad 5^3 = 125, \quad 5^4 = 625, \quad 5^5 = 3125, \quad 5^6 = 15625$$

Exponentiation is much simpler in congruence arithmetic, for we can reduce modulo m at each step. The rules of exponents in ordinary arithmetic become clearer as we practice the simplified computations.

In the following congruence arithmetic use not the least positive residues, as we often do, but the residues that are least in absolute value. For example, modulo 7 use $0, 1, 2, 3, -3, -2, -1$. Then when you square these, you will have $1^2 = (-1)^2 \equiv 1$, $2^2 = (-2)^2 \equiv -3$, etc. (Can you prove that in congruence arithmetic $(-b)^2 \equiv b^2$?)

EXERCISE 7.7. Fill in the powers b^e of integers b in the following table, modulo 7.

power e	integer b						
1	1	2	3	-3	-2	-1	(mod 7)
2	1	-3	2				
3		1	-1	1			
4		2	-3		2	1	
5						-1	
6		1					

EXERCISE 7.8. Make a similar table for congruence modulo 5.

EXERCISE 7.9. Suppose $b^e \equiv 1 \pmod{m}$. Find b^{e+1} and b^{e+2}. Why do we stop with $b^6 \pmod 7$ in Exercise 7.7 and with $b^4 \pmod 5$ in Exercise 7.8?

7.3 BINOMIAL COEFFICIENTS

Our next theorem is about the periodicity of powers modulo a prime, but we first take the opportunity to discuss the binomial coefficients, since they provide us with an elementary way to prove the theorem.

Most of us have made the mistake in elementary algebra of raising a binomial $(a+b)$ to a power, say n, and leaving out the cross-product terms, writing just $a^n + b^n$ instead of the larger (correct) product. Figure 7.2 reminds us of the role the cross-product terms play in the geometry of areas and volumes when $n = 2$ or 3.

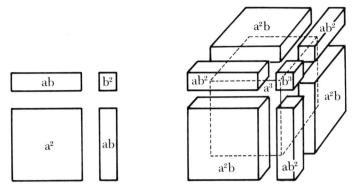

Figure 7.2   Geometric role of cross-product terms in square and cube of a binomial $(a+b)$.

EXERCISE 7.10.   Multiply to find $(a+b)^3$. Find $(3+2)^3$ and compare with $3^3 + 2^3$.

When a binomial $(a+b)$ is raised to the nth power, the integer multipliers C_i^n of terms in a and b are called binomial coefficients.

$$(a+b)^n = \sum_{i=0}^{n} C_i^n a^{n-i} b^i = a^n + C_1^n a^{n-1} b + C_2^n a^{n-2} b^2 + \dots + b^n.$$

To find the value of the binomial coefficient C_i^n, think of $(a+b)^n$ written out as a product of n factors:

$$(a + b)^n = (a + b)(a + b) \dots (a + b).$$

In how many different ways can we get a product $a^{n-i} b^i$? We get this product whenever we use the $+b$ part from i of the factors $(a+b)$ and the a part from the other $n-i$ factors. In how many ways can we select exactly i b's from the n b's available, the other $n-i$ terms contributing a's? The first selection can be any one of the n, followed by the second which can then be any one of the $n-1$ left after the first has been used, followed by the third for which there are left $n-2$ choices, and so on: $n(n-1)(n-2) \dots (n-i+1)$. However, we have counted each of the C_i^n combinations of n b's taken i at a time several times, because we have counted the same combination of b's over again when it is selected in another order. For instance, if we select b from the first 2 factors and a from the rest, we have counted separately $b_1 b_2 a \dots a$ and $b_2 b_1 a \dots a$. In how many ways can a given collection of i b's be ordered? Any one of the i b's can be selected first, any one of the $i-1$ second, and so on, or $i(i-1)(i-2) \dots 2 \cdot 1$. With the notation "$k$ factorial" for $k! = k(k-1)(k-2) \dots 2 \cdot 1$, we can write

$$C_i^n = \frac{n(n-1)(n-2) \cdots (n-i+1)}{i(i-1)(i-2) \cdots 1} = \frac{n!}{i!(n-i)!}$$

An easy way to remember the binomial coefficients is in the form of "Pascal's Triangle," which you can construct stepwise from the relation

$$C_i^{n+1} = C_i^n + C_{i-1}^n. \tag{1}$$

EXERCISE 7.11. Prove relation (1) from the formula for binomial coefficients.

Observe in Figure 7.3 that when n is a prime p, the binomial coefficients C_i^p for $i = 1, 2, ..., p-1$ are all divisible by p.

EXERCISE 7.12. Let p be a prime. From the formula for the binomial coefficients, prove that C_i^p is divisible by p for $i = 1, 2, ..., p-1$.

7.4   FERMAT'S THEOREM AND EULER'S THEOREM

For the following theorem we need the result of Exercise 7.12.

Theorem 7.4 [Fermat's Theorem] *Let b be an integer and p be a prime. Then $b^p \equiv b \,(mod\, p)$. If $p \nmid b$, then $b^{\phi(p)} \equiv 1 \,(mod\, p)$.*

Proof: We make a proof by induction on b. First, to establish the basis of the induction, we notice that the theorem holds for $b = 1$: $1^p = 1 \equiv 1 \,(mod\, p)$, and $1^{\phi(p)} = 1^{p-1} = 1 \equiv 1 \,(mod\, p)$.

C_0^0						1						
C_0^1, C_1^1					1		1					
C_0^2, C_1^2, C_2^2					1	2	1					
C_i^3				1	3	3	1					
C_i^4			1	4	6	4	1					
C_i^5		1	5	10	10	5	1					
	1	6	15	20	15	6	1					
C_i^7	1	7	21	35	35	21	7	1				
	1	8	28	56	70	56	28	8	1			
	1	9	36	84	126	126	84	36	9	1		
	1	10	45	120	210	252	210	120	45	10	1	
C_i^{11}	1	11	55	165	330	462	462	330	165	55	11	1

·   ·   ·   ·   ·   ·   ·   ·   ·   ·   ·   ·

Figure 7.3   Pascal's Triangle

Now take as the induction hypothesis: The theorem holds for $b = s$, so that $s^p \equiv s \pmod p$. Can we establish the result for $s+1$? Using the binomial theorem,

$$(s+1)^p = s^p + C_1^p s^{p-1} + C_2^p s^{p-2} + \ldots + C_{p-1}^p s + 1.$$

But from Exercise 7.12, $C_i^p \equiv 0 \pmod p$ for $i = 1, 2, \ldots, p-1$, so

$$(s+1)^p \equiv s^p + 1 \pmod p.$$

But by the induction hypothesis, $s^p \equiv s \pmod p$. Then

$$(s+1)^p \equiv s + 1 \pmod p,$$

as we needed to show.

As established in Corollary 6.2, an integer b prime to p has an inverse b^{-1} in congruence multiplication modulo p. Then for a b prime to p, we have

$$b^{-1} b^p \equiv b^{-1} b \pmod p,$$

so that

$$b^{\phi(p)} = b^{p-1} \equiv 1 \pmod p. \quad \blacksquare$$

Since by Fermat's Theorem each congruence class satisfies $x^p \equiv x \pmod p$, there is no need to compute powers higher than $p-1$. This idea provides us with a rough check on substitution exercises like the following:

EXAMPLE.   For practice in substitution an algebra student is given a function of x,

$$f(x) = x^5 + x^4 + 2x^3 - 3x^2 + x - 7,$$

and asked to find $f(5)$.

By inspection, $f(5) \equiv f(0) \equiv -7 \equiv 3 \pmod 5$. This gives us a first quick check.

Now consider $f(x)$ modulo 3. From Fermat's Theorem, $x^3 \equiv x \pmod 3$ for each integer (mod 3), so we can write the function

$$f(x) \equiv x^3 \cdot x^2 + x^3 \cdot x + 2x^3 - 3x^2 + x - 7$$
$$\equiv x \cdot x^2 + x \cdot x + 2x - 0 + x - 1 \equiv x^2 + x - 1 \pmod 3.$$

Since $5 \equiv -1 \pmod 3$, we have

$$f(5) \equiv (-1)^2 + (-1) - 1 \equiv 2 \pmod 3.$$

Then the correct answer for $f(5)$ must be congruent to $3 \pmod 5$ and to $2 \pmod 3$, and hence to $8 \pmod{15}$. We could say that a student has only one chance in 15 of satisfying this condition randomly, so an answer congruent to $8\,(\bmod\,15)$ is probably correct!

EXERCISE 7.13.   A student is given a function of x,

$$f(x) = x^5 - 2x^4 - 3x^3 + x^2 + 1,$$

and asked to compute $f(4)$.

Find $f(4)$ $(\bmod\,4)$ and $(\bmod\,3)$, and hence $(\bmod\,12)$.

In theorem 7.3 we learned that the function $\phi(m)$ is a multiplicative function; that is, the function of a product is the product of the function values, provided that the factors are relatively prime. We can use this fact to generalize Fermat's Theorem:

Theorem 7.5 [Euler's Theorem]   *Let c be an integer prime to a positive integer m. Then*

$$c^{\phi(m)} \equiv 1\,(mod\,m)$$

Proof: First we establish by induction on the exponent a that the theorem holds for a prime-power modulus p^a. Then the theorem follows from the multiplicative nature of $\phi(m)$. Let p^a, where $a \geqslant 1$, be a power of a prime p. *Basis:* For $a = 1$, Theorem 7.4 provides the relation $c^{\phi(p)} \equiv 1\,(\bmod\,p)$. Suppose as induction hypothesis that for a certain $a \geqslant 1$ we know that

$$c^{\phi(p^a)} \equiv 1\,(\bmod\,p^a).$$

We rewrite this according to the definition of congruence as

$$c^{\phi(p^a)} = 1 + qp^a,$$

and note that $\phi(p^{a+1}) = p\phi(p^a)$.

By the binomial theorem we can find

$$c^{\phi(p^{a+1})} = c^{p\phi(p^a)} = [c^{\phi(p^a)}]^p = [1+qp^a]^p$$

$$= 1^p + p(1)^{p-1}(qp^a) + \frac{p(p-1)}{2}(1)^{p-2}(qp^a)^2 + \ldots$$

$$\equiv 1\,(\bmod\,p^{a+1}),$$

completing the induction.

Now let c be prime to a modulus $m = p_1^{a_1} p_2^{a_2} \dots p_k^{a_k}$. Since $c^{\phi(m)} - 1$ is divisible by $c^{\phi(p^a)} - 1$ for each of the prime-power divisors of m, which in turn is divisible by p^a, we have $c^{\phi(m)} - 1 \equiv 0 \pmod{m}$, or

$$c^{\phi(m)} \equiv 1 \pmod{m}. \quad \blacksquare$$

EXERCISE 7.14.   The end of the last proof hinged on the fact that if $r = st$, then $c^r - 1$ is divisible by $c^s - 1$. Prove this fact by calculating the quotient.

EXERCISE 7.15.   Consult Figure 7.1 for $\phi(15)$. Show that $c^{\phi(15)} \equiv 1 \pmod{15}$ for each positive c less than 15 and prime to 15. What is the minimal exponent e for which $c^e \equiv 1$ in each case?

Points to Remember

1.   The Euler phi function $\phi(n)$ gives the number of different integers relatively prime to $n \pmod{n}$.

2.   The function $\phi(p) = p - 1$, $\phi(p^w) = p^{w-1}(p-1)$ for a prime power p^w, and the function is multiplicative.

3.   If c is relatively prime to m, then $c^{\phi(m)} \equiv 1 \pmod{m}$ (Euler's Theorem).

SUGGESTED PROJECT

A talk or report can be built around Pascal's Triangle. In the course of a demonstration the audience can help generate later rows of the Triangle by adding adjacent numbers from the previous row,

$$C_i^{n+1} = C_i^n + C_{i-1}^n.$$

Show how to find powers of a binomial. Physical models based on Figure 7.2 might be helpful. Show by inspection that in the row of Pascal's Triangle for a prime p, p divides all the coefficients except the bordering 1's.

CHAPTER 8

Primitive Roots and Indices

8.1 ORDER OF AN ELEMENT

As you can see from Euler's Theorem 7.5, every integer b prime to m can be raised to some positive exponent to give 1 modulo m. (In the theorem the exponent $\phi(m)$ is used for all b.) We are going to give a name to the smallest such exponent for a given b (Definition 8.1 of *order*) and then discuss integers for which that exponent is maximal. Such an integer can be raised to different powers, called *indices*, to give all the integers (mod m) that are prime to m.

Definition 8.1 *Let* $(b, m) = 1$. *The* **order** *of b modulo m is the smallest positive integer e for which*

$$b^e \equiv 1 \, (mod \, m) \, .$$

We say that b **belongs** *to the exponent e (mod m).*

Does every integer b that is prime to m have an order? We can use Euler's Theorem to show that it does:

Theorem 8.1 *If* $(b, m) = 1$, *then b has an order* $e \, (mod \, m)$ *and e divides every positive integer f for which* $b^f \equiv 1 \, (mod \, m)$; *in particular,* $e \, | \, \phi(m)$.

Proof: From Theorem 7.5, since b is prime to m, $\phi(m)$ provides one example of a positive exponent f for which $b^f \equiv 1 \, (mod \, m)$, so we can choose e to be the least positive one.

Now suppose f is any positive integer for which

$$b^f \equiv 1 \, (mod \, m).$$

Use the division algorithm (Theorem 2.2) to write

$$f = qe + r, \qquad 0 \leqslant r < e.$$

Then $b^f = b^{qe+r} = b^{qe}\, b^r = (b^e)^q\, b^r \equiv 1^q\, b^r = b^r \pmod{m}$. But $b^f \equiv 1 \pmod{m}$ and $r < e$, with $b^r \equiv b^f \equiv 1$. Since e is minimal, we conclude that $r = 0$, so that $e \mid f$.

Since $\phi(m)$ constitutes an f for which $b^f \equiv 1$, we have $e \mid \phi(m)$. ∎

EXAMPLE.   Modulo 12, 1 has order 1, 5 has order 2, 7 has order 2, and 11 has order 2. If $5^f \equiv 1 \pmod{12}$, then 2, the order of 5, divides f; in particular, 2 divides $\phi(12) = 4$.

The distinct powers of an element b prime to the modulus m supply another example of congruence arithmetic, this time an arithmetic of exponents taken modulo e, the order of b modulo m.

EXERCISE 8.1.   Verify Theorem 8.1 for $b = 2$ and $m = 15$.

EXERCISE 8.2.   Find the order of $3 \pmod{7}$ and show that it equals $\phi(7)$. Write the distinct powers of $3 \pmod{7}$ and compare them with the positive integers less than 7.

Definition 8.2   *A* **primitive root** *modulo m is an integer u whose order (mod m) is $\phi(m)$.*

From Theorem 8.1, each integer b for which $(b, m) = 1$ has an order e that is a divisor of $\phi(m)$. The primitive roots, then, are those whose orders not only divide but equal $\phi(m)$. Not every modulus has primitive roots.

Theorem 8.2   *A primitive root (mod m) exists for $m = 2$, 4, p^a, and $2p^a$, where p is an odd prime and a is a positive integer. There is no primitive root for other values of m.*

For some simple cases, we find

$\phi(2) = 1$, so that 1 is primitive $\pmod{2}$.

$\phi(3) = 2$, and 2 belongs to $2 \pmod{3}$, and hence is primitive.

$\phi(4) = 2(1) = 2$, and 3 belongs to $2 \pmod{4}$; hence it is primitive.

$\phi(5) = 4$, and 2 belongs to $4 \pmod{5}$.

$\phi(6) = (1)(2) = 2$, and 5 belongs to $2 \pmod{6}$.

$\phi(7) = 6$, and 3 belongs to $6 \pmod{7}$.

$\phi(8) = 4(1) = 4$, but $1 \equiv 1$, $3^2 = 1$, $5^2 \equiv 1$, and $7^2 \equiv 1$, so there is no primitive root modulo 8.

We shall not build up the sequence of theorems necessary to prove Theorem 8.2 here.* It is interesting to note that although the *existence* of a primitive root modulo a prime p can be proved, there is no construction or algorithm for finding one, just as there is no construction available for prime integers. It is fairly easy to show how to construct a primitive root modulo p^a from a primitive root modulo p and a primitive root modulo $2p^a$ from that.

8.2   INDICES

Suppose u is a primitive root (mod p). The powers

$$u, u^2, u^3, ..., u^{p-1} \equiv 1$$

of u are distinct (mod p), because $u^s \equiv u^t$ with $0 < s < t < p$ would imply that

$$u^{p-1-s} u^s \equiv u^{p-1-s} u^t,$$

so that

$$1 \equiv u^{t-s}.$$

Then by Theorem 8.1, the order of u, $p-1$, would divide $t-s$, but $0 < t-s < p-1$.

Since $u, u^2, u^3, ..., u^{p-1} \equiv 1$ are all incongruent (mod p), they are, in some order, the $p-1$ integers not congruent to zero (mod p). Then the exponents can be used for computing, just as we use logarithms. The exponents are called **indices to base u modulo p**.

EXERCISE 8.3.   Let the modulus $m = p$ be 29, and the primitive root u(mod 29) be 2. Complete this list of powers of 2 reduced (mod 29) to integers in the range from -14 through $+14$, writing the powers $2^0, 2^1, ..., 2^{27}$ in two long vertical columns on ruled paper or in typed form, so as to have equal spacing (Fig. 8.1).

Cut the two columns apart. Now to multiply, say, -13 and 9(mod 29), think of adding exponents in $(-13)(9) \equiv (2^4)(2^{10}) \equiv 2^{14} \equiv -1$. Check that $(-13)(9) \equiv -1$(mod 29). Now accomplish this multiplication-by-adding operation by setting the "1" of the C scale opposite the "-13" of the D scale, sliding one scale along the cut so the two scales remain parallel (Fig. 8.2). Opposite "9" on the C scale read off the product "-1" on the D scale. As you see, you have made a slide rule operating on the principle of adding indices to multiply numbers, just as an ordinary slide rule adds logarithms to multiply numbers.

---

* Such a sequence can be found in I. Niven and H. S. Zuckerman, *An Introduction to the Theory of Numbers*, John Wiley and Sons, 1960, Sections 2.7 and 2.9 and Problems 19 and 20, page 52.

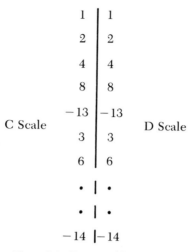

C Scale   D Scale

Figure 8.1   Powers of 2 modulo 29.

EXERCISE 8.4.   Find by experiment a primitive root (mod 7) and construct a mod 7 slide rule as in the previous exercise.

We often use 1/7 as an example in arithmetic for representing a fraction by a repeating decimal, because it has a long (6-digit) period (Fig. 8.3).

The fact is that $10 \equiv 3 \pmod 7$ has order $7 - 1 = 6$ and so is primitive (mod 7). This means that 6 is the smallest power of 10 that leaves a remainder of 1 upon division by 7 and so starts the cyclic division over again with 7/1.

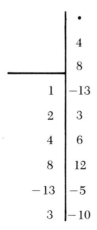

Figure 8.2   Slide rule in position to multiply by $-13 \pmod{29}$.

$$0.\overline{142857}$$
$$7\overline{\smash{)}1.000000}$$
$$\underline{7}$$
$$30$$
$$\underline{28}$$
$$20$$
$$\underline{14}$$
$$60$$
$$\underline{56}$$
$$40$$
$$\underline{35}$$
$$50$$
$$\underline{49}$$
$$1$$

Figure 8.3   Periodic decimal with period 6.

Notice that $10 \equiv 1 \pmod 3$ has order 1 and $1/3 = 0.\overline{3}$. Also, $10 \equiv -1 \pmod{11}$ has order 2 and $1/11 = 0.\overline{09}$. Since 10 has order $6 \pmod{21}$, the period length of $1/21$ is not $\phi(21) = 2 \cdot 6 = 12$, but only 6 digits: $1/21 = 0.\overline{047619}$, as we can see from $1/21 = (1/7)/3$.

Once we have a primitive root modulo p we can find any other primitive roots from it and also find which integers belong to smaller divisors of $p-1$. We illustrate the method for $p = 11$ and the primitive root $u = 2$ in Figure 8.4.

$$2^1, \quad 2^2, \quad 2^3, \quad 2^4, \quad 2^5, \quad 2^6, \quad 2^7, \quad 2^8, \quad 2^9, \quad 2^{10},$$

are congruent $\pmod{11}$ to

$$2, \quad 4, \quad -3, \quad 5, -1, \quad -2, -4, \quad 3, -5, \quad 1 .$$

Since $2^{10} \equiv 1$, 2^2, 2^4, 2^6, and 2^8 have order 5; that is, 2^f, where $(f, 10) = 2$. There are $\phi(5) = 4$ of these, for finding the integers among $2, 4, 6, 8, 10$ such that $(f, 10) = 2$ is like finding the integers among $1, 2, 3, 4, 5$ with $(i, 5) = 1$.

Similarly, there is $\phi(2) = 1$ integer, 2^5, that belongs to $2 \pmod{11}$. There is $\phi(1) = 1$ integer, 2^{10}, that belongs to $1 \pmod{11}$. All the others, 2^1, 2^3, 2^7, and 2^9 belong to 10 and so are primitive. There are $\phi(10) = 4$ of them, as they can be found by raising 2 to those powers f less than 10 with $(f, 10) = 1$.

Figure 8.4   Orders of the powers of 2 modulo 11.

EXERCISE 8.5.   Verify in Fig. 8.4 that $10 = \sum_{d|10} \phi(d)$; that is, that 10 is equal to the sum of $\phi(d)$ taken over all positive divisors d of 10. Show why for any prime p we have $p-1 = \sum_{d|p-1} \phi(d)$.

EXERCISE 8.6.   Let $d|m$. Show that $\phi(d)$ of the positive residues modulo m, $1, 2, 3, ..., m$, have the property that $(x, m) = m/d$.

EXERCISE 8.7.   Use the result of Exercise 8.6 to extend the result of Exercise 8.5 to any positive integer m:

$$m = \sum_{d|m} \phi(d).$$

8.3 POWER RESIDUES

If the modulus is an odd prime p, we know there is a primitive root u and that each positive residue (mod p) can be written as a power of u. The residues that are written as even powers of u, that is, have even index (mod p) to base u, are called quadratic residues. They are the residues of perfect squares (mod p).

An integer n that is prime to a modulus m is called a kth power residue (mod m) if it is congruent to a perfect kth power of some integer b:

$$n \equiv b^k (\text{mod } m), \quad \text{and} \quad (n, m) = 1.$$

EXAMPLE.   Modulo 11,

$$1^2 \equiv 10^2 \equiv 1$$
$$2^2 \equiv 9^2 \equiv 4$$
$$3^2 \equiv 8^2 \equiv 9$$
$$4^2 \equiv 7^2 \equiv 5$$
$$5^2 \equiv 6^2 \equiv 3$$

Then the quadratic residues modulo 11 are 1, 4, 9, 5, and 3. All the non-zero integers are 3rd power residues modulo 11, for $1^3 \equiv 1$, $2^3 \equiv 8$, $3^3 \equiv 5$, $4^3 \equiv 9$, $5^3 \equiv 4$, $6^3 \equiv 7$, $7^3 \equiv 2$, $8^3 \equiv 6$, $9^3 \equiv 3$, and $10^3 \equiv 10$.

Modulo 9, 3 is not a kth power residue, since $(3, 9) = 3 \neq 1$.

EXERCISE 8.8.   Show that modulo 11 all the quadratic residues are also 4th power residues.

EXERCISE 8.9.   Show that there are exactly two 5th power residues modulo 11.

Some famous theorems of number theory are concerned with power residues, notably the Quadratic Reciprocity Theorem about quadratic residues. The quadratic reciprocity theorem has an aesthetic appeal for many who appreciate elegance in mathematical deduction. Although the proofs are rather complicated, they are accessible to you at this point with a little extra reading. [See Chapter 3 of I. Niven and H. S. Zuckerman, *An Introduction to the Theory of Numbers*, John Wiley, and Sons 1966.]

8.4 HIGHER DEGREE CONGRUENCES

The kth power residues modulo m are the residues a for which the congruence

$$x^k \equiv a \,(\mathrm{mod}\, m) \qquad \text{and} \qquad (a, m) = 1$$

has solutions. This is a congruence of degree k. We have covered the case of linear congruences completely (Theorems 6.1 and 6.2). Some theory is available for more general congruences of higher degree

$$a_k x^k + a_{k-1} x^{k-1} + \cdots + a_1 x + a_0 \equiv 0 \,(\mathrm{mod}\, m).$$

We noticed in Exercise 6.1 one of the peculiarities of higher degree congruences with a composite modulus.

Point to Remember

1. For a prime modulus p there is a primitive root u that can be raised to indices $1, 2, \ldots, p-1$ to give the non-zero integers $(\mathrm{mod}\, p)$ in some order.

SUGGESTED PROJECT

Exponentiation, primitive roots, and indices offer material for a report, talk, or demonstration. To keep the presentation elementary, restrict it to numerical examples. Experimentation will convince the audience that in congruence arithmetic the increasing powers of an integer prime to the modulus eventually produce 1. For instance, modulo 10, $3^1 \equiv 3$, $3^2 \equiv 9$, $3^3 \equiv 3 \cdot 3^2 \equiv 7$, and $3^4 \equiv 3 \cdot 7 \equiv 1$. Let the audience practice exponentiation for a while to help them recall that 3^4 means $3 \cdot 3^3$, for instance. Ordinary arithmetic gives us little chance to practice raising numbers to powers, since the increase in size is so rapid.

Next, show some special cases of primitive roots for prime moduli, write their distinct powers, and show that they provide all the integers prime to the modulus. Multiply integers modulo p by writing them as powers of a primitive root and adding exponents, as in Exercise 8.3.

Demonstrate the use of a modular slide rule like the one described in Exercise 8.3. Explain how an ordinary slide rule multiplies numbers by adding lengths that correspond to exponents applied to a base of 10. Compare the base 10 to the primitive root modulo p.

CHAPTER 9

Diophantine Problems

9.1 LINEAR PROBLEMS

EXAMPLE 1. A student's transcript shows t 3-hour courses and f 5-hour courses, for a total of 64 hours. Find t and f. From the description of the problem we get the equation

$$3t + 5f = 64.$$

This problem and the equation it gives rise to are called **Diophantine**. They show the typical characteristics: (1) The equation by itself is indeterminate; it may have many solution sets (t,f). (2) The conditions of the problem supply additional restrictions on the solutions: in the case of Example 1, as in many Diophantine problems, we have the restrictions that the solutions must be integral and that they must be positive, or at least non-negative.

Actually, Diophantus did not treat simple linear problems like our example, but nowadays we use his name for indeterminate problems with special restrictions, especially restriction to integral or rational solutions. Diophantus lived at Alexandria perhaps sometime around A.D. 250, although his few surviving mathematics books are almost all we know of him. At that time Alexandria was under Roman sway, the Romans having subjugated the Ptolemies of Egypt.

Situated on the African shore of the Mediterranean Sea, Alexandria benefitted from the busy commerce. Under the Ptolemies Alexandria had become a center of learning, with its famous Library and a Museum somewhat akin to a university. By the third century Greek science and philosophy, mathematics included, had declined from its height, having produced major contributions in geometry with much less emphasis on algebra. Although Diophantus is considered to have been a writer of Greek algebra, there may well have been influences from Babylon or India by the time he wrote.

Some classes of Diophantine problems can be solved in general, but due to the wide range of possible restrictions, many of them must be analyzed individually, sometimes very ingeniously. These problems fascinate those who like puzzles. Some of the very puzzles proposed by Diophantus still pop up from time to time.

To solve Example 1 experimentally, note that f must be less than 13, no matter what t is, since $5 \cdot 13 = 65 > 64$. Try $f = 12$. This would give us $3t + 5(12) = 64$, or $3t = 4$, which has no integral solution for t. Continuing to decrease f, we find 4 possible solution-pairs, corresponding to the 4 lattice-points in the first quadrant lying on the line $3t + 5f = 64$ in Figure 9.1.

EXAMPLE 2.   On a "right minus wrong" test, a student gets 10 points for each of r right answers, but loses 2 points (gets -2) for each of w wrong answers. If his net score was 85, find r and w.

We have $10r - 2w = 85$, or $2(5r - w) = 85$, but $2 \nmid 85$, so there are no solutions.

Theorem 9.1   *An equation $ax + by = c$ has integral solutions (x_0, y_0) if and only if $(a, b) \mid c$.*

Proof: Let (x_0, y_0) be a solution pair. This means that

$$ax_0 + by_0 = c.$$

Let d be a common divisor of a and b, so that $a = da'$ and $b = db'$. Then

$$ax_0 + by_0 = da'x_0 + db'y_0 = d(a'x_0 + b'y_0) = c, \quad \text{so that} \quad d \mid c.$$

Since (a, b) is a common divisor d, (a, b) must divide c.

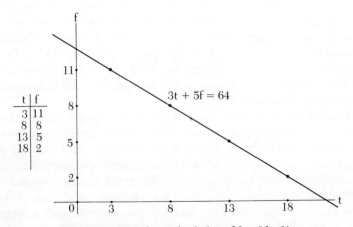

Figure 9.1   Experimental solution of $3t + 5f = 64$.

Conversely, suppose $(a, b) \mid c$, so that $c = c'(a, b)$. From Theorem 2.3 there are integers r and s for which

$$ra + sb = (a, b).$$

Multiplying by c' gives us

$$(c'r)a + (c's)b = c'(a, b) = c.$$

Then $(x, y) = (c'r, c's)$ is a solution in integers for $ax + by = c$. ∎

EXERCISE 9.1.   Write $ax + by = c$ as a congruence modulo b, and use Theorem 6.1 to prove that the congruence has no solution if $(a, b) \nmid c$. Argue that hence the equation has no solution.

Continuing Example 1, we can take $r = 2$ and $s = -1$, since $2 \cdot 3 + (-1)5 = 1 = (3, 5)$. We have $c' = 64$, because $64 = (3, 5)64$. From the proof just above $(x, y) = (c'r, c's) = (64 \cdot 2, 64(-1))$ should be an integral solution, and we can verify that it is.

In this case the particular solution we have found for the Diophantine equation $3x + 5y = 64$ is not a solution for the original problem. It is integral, as needed, but not positive.

Theorem 9.2 *Let $ax + by = c$ be a Diophantine equation with $a = a'(a, b)$ and $b = b'(a, b)$, and let (x_0, y_0) be a solution. Then every solution (x_i, y_i) is of the form $(x_0 + b'n, y_0 - a'n)$, where n is an integer.*

Proof: First, check to see that $(x_0 + b'n, y_0 - a'n)$ is a solution for the equation.

$$a(x_0 + b'n) + b(y_0 - a'n) = ax_0 + ab'n + by_0 - ba'n$$
$$= (ax_0 + by_0) + [(a, b)a'b'n - (a, b)b'a'n]$$
$$= c + 0 = c.$$

Next, let $(x_0 + h, y_0 + k)$ stand for any solution. Then $a(x_0 + h) + b(y_0 + k) = c = ax_0 + by_0$. We can cancel like terms to get

$$ah = -bk,$$

or, dividing by (a, b),

$$a'h = -b'k.$$

Since $(a', b') = 1$, we have $b' \mid h$, so that $h = b'n$ for some integer n. Then $k = -a'n$. ∎

In a problem like that of Example 1, where positive solutions are required, we apply Theorem 9.2 to the particular solution Theorem 9.1 gives us. In Example 1, Theorem 9.1 gave us $(128, -64)$. In that problem $a = a' = 3$. We want $y_0 - a' n = -64 - 3n > 0$, so we take $n < -21$. For $n = -22$, $(x_0 + b' n, y_0 + a' n) = (128 + 5(-22), -64 - 3(-22)) = (18, 2)$, one of the positive solutions we found experimentally.

EXERCISE 9.2.   On a machine-graded test a student gets x 5-point questions right and y 10-point questions right with a total grade of 97. He thinks something is wrong. Why?

EXERCISE 9.3.   In the stock market John trades (buys *or* sells) x shares of Canadian oil stock at 6 cents a share and trades y shares of mining stock at 14 cents a share, for a net loss of 4 cents. Find a solution for (x, y). Find a solution in case the total number of shares that changed hands was 286.

EXERCISE 9.4.   A baseball pitcher strikes out s players on 3 pitches each and walks w players on 4 pitches each, for a total of 17 pitches. Find s and w. How many possible solutions has this problem?

EXERCISE 9.5.   For a father-son banquet, a caterer supplies 8 oz. of meat for each man and 6 oz. for each boy, for a total of 13 lb. 2 oz. How many guests were there? [This exercise contains a special requirement that favors one from a wide field of solutions for the indeterminate equation.]

EXERCISE 9.6.   A shopper buys x 2-cent items, y 6-cent items, and z 10-cent items. Show that the total bill cannot be $1.77. Find a solution-triple in case the total is $1.98. Invent more detail for the problem so as to favor a solution with $x > y > z$.

9.2   PYTHAGOREAN TRIPLES

The integers 3, 4, 5 have the property that $3^2 + 4^2 = 5^2$. Geometrically, this means that these integers can serve as the sides of a right triangle with 5 as the hypotenuse. The Egyptians used this principle to construct perpendiculars, as in Figure 9.2.

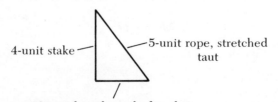

4-unit stake   5-unit rope, stretched taut

3-unit length marked on base

Figure 9.2 Pythagorean construction of a perpendicular.

Definition 9.1   *A* **Pythagorean triple** *is a set of three integers* x, y, z *for which* $x^2 + y^2 = z^2$. *The triple is* **primitive** *if* x, y, z *have no common prime divisor.*

EXERCISE 9.7.   Show that $(6, 8, 10)$ is a Pythagorean triple. Show that $(9, 12, 15)$ is a Pythagorean triple. Let (x_0, y_0, z_0) be a Pythagorean triple; prove that for any positive integer n, (nx_0, ny_0, nz_0) is a Pythagorean triple.

EXERCISE 9.8.   One Pythagorean triple is $(10, 24, 26)$. Find the related primitive triple.

The search for primitive Pythagorean triples is a Diophantine problem: We need solutions (x, y, z) for the indeterminate equation $x^2 + y^2 = z^2$ with the restrictions that x, y, z be positive integers with no common prime factor. We can build up a profile of a typical primitive triple.

First, consider the parity of x and of y: that is, whether they are odd or even. Suppose both are odd. For some integers k and m, we would have $x = 2k+1$, $y = 2m+1$. Then $x^2 + y^2 = z^2 = (2k+1)^2 + (2m+1)^2 = 4(k^2 + k + m^2 + m) + 2$. But there is no integer with a square like this, for if z is even, z^2 is divisible by 4; if z is odd, z^2 is not divisible by 2.

EXERCISE 9.9.   Recast this analysis in terms of congruence modulo 4. Which of the residues $(\bmod 4)$, $0, 1, 2, 3$, are quadratic residues, that is, which are congruent $(\bmod 4)$ to perfect squares?

EXERCISE 9.10.   Suppose x and y are both even. Then show that $2 \mid z$, so that x, y, z cannot be primitive.

Since it does not matter which of the smaller integers we call x and which we call y, we agree to call x the odd one and y the even one.

Continuing our investigation, we notice that since x, y, and z are positive integers, there exist positive integers r and s with $(r, s) = 1$ for which

$$y = (x + z)s/r.$$

From this we get a formula for x in terms of y and z, which we substitute for x in the equation $x^2 + y^2 = z^2$:

$$x = \frac{yr}{s} - z.$$

$$\left(\frac{yr}{s} - z\right)^2 + y^2 = z^2,$$

or, multiplying by s^2,

$$r^2 y^2 - 2rsyz + s^2 z^2 + s^2 y^2 = s^2 z^2,$$

so either $y = 0$ or

$$y = z2rs/(r^2 + s^2).$$

Then x in terms of z is $x = (r/s)(z2rs/(r^2 + s^2)) - z$, or

$$x = z(r^2 - s^2)/(r^2 + s^2).$$

Then the triple x, y, z must be proportional to

$$(r^2 - s^2)/(r^2 + s^2), \qquad 2rs/(r^2 + s^2), \qquad 1;$$

or multiplying by $r^2 + s^2$, we have the proportion

$$x:y:z = r^2 - s^2 : 2rs : r^2 + s^2.$$

We have $r > s$, and $r \not\equiv s \pmod 2$. The first of these restrictions is made because r must be positive; the second, because r must be odd (see Exercise 9.14).

EXERCISE 9.11.   We have shown that every primitive Pythagorean triple stands in a proportion $x:y:z = r^2 - s^2 : 2rs : r^2 + s^2$, where $r > s$ are integers with $r \not\equiv s \pmod 2$. Prove the converse, by showing that any positive integers r and s with $(r, s) = 1$ and $r > s$ and $r \not\equiv s \pmod 2$ can be used in the proportion formula to generate a Pythagorean triple. Is the triple primitive?

EXERCISE 9.12.   Find r and s for the triples $(3, 4, 5)$ and $(5, 12, 13)$. Use other r, s pairs to construct two other primitive Pythagorean triples.

9.3   DIFFERENCE OF TWO SQUARES

Consider this Diophantine problem: Which positive integers n can be written as the difference between two squares?

$$n = x^2 - y^2.$$

EXERCISE 9.13.   Show by experiment that 6 and 10 cannot be represented in the form $x^2 - y^2$.

EXERCISE 9.14.   Study the possible values for $n \pmod 4$ in case x and y are both even; in case x is even and y is odd; in case x is odd and y is even; in case x and y are both odd. Which value is missing for n? Compare with Exercise 9.13.

EXERCISE 9.15. Prove that for any odd prime p there is a solution (x, y) for $p = x^2 - y^2$. [Write $p = x^2 - y^2 = (x+y)(x-y)$. Since p is a prime $x - y$ must equal ?] Find a solution-pair for $p = 13$.

EXERCISE 9.16. Prove that if $n = a \cdot b$ with a and b both odd, then $n = x^2 - y^2$ has a solution with $x = (a+b)/2$. Use this to solve the special case $n = p$, an odd prime. (Cf. Exercise 9.15.)

EXERCISE 9.17 If $n = 2^{2e}m$ and $m = x_0^2 - y_0^2$, find a solution-pair (x, y) for $n = x^2 - y^2$.

EXERCISE 9.18. Combine the results of the last four exercises to deduce exactly which positive integers n can be expressed as the difference between two squares.

9.4 THE FOUR-SQUARE THEOREM

Diophantus used, apparently without proof except for observation, the fact that every positive integer can be expressed as the sum of 4 squared integers. In this case some of the squares can be 0^2, so that we could say alternatively that each integer can be expressed as the sum of 4 or fewer positive squares. Fermat, much of whose number theory work survives in the margins of his copy of a translation of Diophantus' work, said he had been able to prove the 4-square theorem, but no copy of this proof survives if indeed Fermat wrote it down. The great Euler published in 1751 some results he had proved in 20 years of work on the problem, but could not at that time prove the whole 4-square theorem. The first complete proof published was by Lagrange in 1770, and the author acknowledged the help Euler's paper had given him. Then in 1773 Euler was able to offer a much simpler proof.

To round out this digression on history, let us record that the four-square theorem is known as "Bachet's Theorem" after Claude-Gaspar Bachet, Sieur de Meziriac (1581–1638). In 1612 Bachet published a collection of number puzzles that was very popular and much reprinted. In 1621 he published an edition of *Diophantus*, not the first, but by far the best. It was a copy of Bachet's *Diophantus* that served Fermat as a notebook and provided the margin for notes that still have mathematicians guessing. [Although Bachet's theorem has been proved, Fermat's famous "last theorem," which we shall discuss later, has not, despite Fermat's intriguing and frustrating note in the margin of Bachet that he had a proof too long for the margin!]

As you see, the transition from Diophantus to Fermat was surprisingly immediate. Some number theory puzzles have attracted people across a wide span of time and place, people of all occupations. Perhaps you will find yourself in the long parade of contributors.

EXERCISE 9.19. Show that 7 cannot be represented as the sum of one square, two squares, or three squares. This proves that the "4" in the 4-square theorem is minimal.

Many numbers can be expressed as a sum of fewer than 4 squares (or with some of the four equal to 0^2). In fact, the only integers not expressible as the sum of three squares (with 0^2 still permitted) are integers of the form $4^e(8m+7)$.

Which integers can be expressed as the sum of fewer than 3 squares? It can be proved that n can be expressed as the sum of two squares unless its factorization into prime powers contains a factor p^e where p is a prime of the form $4m+3$ and e is odd.

EXERCISE 9.20. Write the first 50 positive integers as sums of as few squares as possible, comparing with the results given above about sums of 4, 3, and 2 squares.

EXERCISE 9.21. Verify Euler's Lemma that

$$(a^2+b^2+c^2+d^2)(x^2+y^2+z^2+w^2)$$
$$= (ax+by+cz+dw)^2 + (ay-bx+cw-dz)^2$$
$$+ (az-cx+dy-bw)^2 + (aw-dx+bz-cy)^2.$$

Write 1 and 2 as sums of four squares. Put this information together to prove that if every odd prime can be expressed as a sum of four squares, then every positive integer can.

EXERCISE 9.22. Read the proof of Bachet's Theorem in I. Niven and H. S. Zuckerman, *Introduction to the Theory of Numbers*, John Wiley and Sons, 1966, pp. 114–115.

Theory of numbers is not noted for practical application, but occasionally it is put to some use outside the realm of puzzles. Certainly Bachet's four-square theorem seems an unlikely candidate for application. Who would want to write an integer as the sum of four squares, anyway? Statisticians would. In a paper in *Annals of Mathematical Statistics* (vol. 26, No. 1, "Note on Linear Hypotheses with Prescribed Matrix of Normal Equations") John Maxfield and R. S. Gardner applied Bachet's theorem to a problem in designing statistical experiments, demonstrating once again that what is "applicable" depends on what techniques are available and how great the need is to solve a problem.

9.5   WARING'S PROBLEM

In 1770, the same year that saw publication of Lagrange's proof of Bachet's Theorem, the English mathematician Waring published his conjecture from numerical experimentation that there is some number $g(3)$ for which it can be stated "Every positive integer n can be expressed as the sum of $g(3)$ cubes," some number $g(4)$ for which "Every n is the sum of $g(4)$ fourth powers," and in general some number $g(k)$ for which "Every n is the sum of $g(k)$ kth powers." Hilbert was able to prove that such a $g(k)$ does exist for each k. As we have reported, $g(2) = 4$. It has been proved that $g(3) = 9$. It is known that $g(4)$ is somewhere in the range $19 \leqslant g(4) \leqslant 35$ and that $37 \leqslant g(5) \leqslant 54$. For $k \geqslant 6$, $g(k)$ has been evaluated [with a possible exception]. For $k > 3$ the known proofs are part of "analytic number theory" where the techniques of limits (as in calculus) are used.

EXERCISE 9.23.   Consult L. E. Dickson's famous *History of the Theory of Numbers*, Volume II, Chelsea, 1952, Preface, pages ii and iii, about the problem of sums of m-gonal numbers, a generalization of Bachet's problem. Page 1 shows an illustration of triangular, square, and pentagonal numbers. While you have the volume in hand, look over the references and descriptions in Chapters 6 to 8 on sums of 2, 3, and 4 squares.

SUGGESTED PROJECTS

1.   Linear Diophantine problems have a special appeal, because they are easy to understand and because there are lots of right answers. The audience can participate in the fiction of the problem by contributing extra conditions that favor one or another of the possible answers. An experimental approach is especially valuable, as it encourages analysis of what the problem means. The audience can be helped to discover that some Diophantine problems have no solution.

2.   A short talk can be presented on the subject of Pythagorean triples. A poster showing some triples might be used to lead to questions about how to generate them. The Egyptian construction for a perpendicular shown in Figure 9.2 makes a good demonstration or physical model.

CHAPTER 10

Fibonacci Numbers

The only European who was outstanding for mathematical activity during the Middle Ages was Leonardo Fibonacci of Pisa, or Leonardo Pisano. [In doing reference work check under Fibonacci, Leonardo, and Pisano.] He was one of the men of learning collected by the patron Emperor Frederic II as ornaments for his court. Fibonacci's father was a Pisan who had business interests in North Africa, and the son was taught by a Muslim tutor and traveled in Egypt, Greece, and Syria. He received the legacy of algebra from the Arab world, which in turn incorporated whole passages from the works of Diophantus. In his main work, *Liber Abaci*, which came out in 1202, he used many problems translated word for word from the main work of al-Karkhi of Bagdad (died about 1030), in so doing re-copying some material al-Karkhi had translated word for word from Diophantus. Here we have one of the curious short-circuits by which number theory has been shunted across time and distance.

10.1 THE FIBONACCI SEQUENCE

The "Fibonacci sequence" arose in connection with a puzzle Fibonacci proposed about rabbits in his *Liber Abaci*:

"How many pairs of rabbits will be produced in a year, beginning with a single pair, if in every month each pair bears a new pair which becomes productive from the second month on?"

We might try a solution by a table such as that in Figure 10.1.

Notice that each month the baby pairs grow up and are replaced by adult pairs, making the new "adult" entry the previous one plus the previous "baby" entry. Each of the pairs that was adult the last month produces one baby pair, so the new "baby" entry equals the previous "adult" entry.

	start	after 1 mo.	after 2 mos.	3	4	5	6	7	8
adult pairs	1	1	2	3	5	8	13	21	34
baby pairs	0	1	1	2	3	5	8	13	21
No. of pairs	1	2	3	5	8	13	21	34	55

Figure 10.1   Chart of rabbit population by month.

Except for starting point, the three horizontal rows of the table are alike, and we name them the **Fibonacci sequence,** following the example of the 19th century French mathematician, E. Lucas.

$$F_0 = 0, \; F_1 = 1, \; F_2 = 1, \; F_3 = 2, \; F_4 = 3, \; F_5 = 5, \; \ldots$$

From the rather idealized habits of rabbits in the generating example, we have a recursion, or step-wise, formula for generating later Fibonacci numbers:

$$F_{n+1} = F_n + F_{n-1}; \qquad n > 0, \; F_1 = 1, \; F_2 = 1. \tag{1}$$

This kind of a recursion formula can be generalized to give a so-called **generalized Fibonacci sequence**

$$u_{n+1} = u_n + u_{n-1}; \qquad u_1, \, u_2 \text{ to be chosen.} \tag{2}$$

EXERCISE 10.1.   The generalized Fibonacci sequence with $u_1 = 1$, $u_2 = 3$ has been called the **Lucas sequence** after the mathematician who named the Fibonacci sequence. Write the first dozen Lucas numbers. Show how to adapt the rabbit problem so as to obtain the Lucas numbers and find the total number of rabbits at the end of one year in this case.

Let $\phi = (1 + \sqrt{5})/2$, and $\bar{\phi} = (1 - \sqrt{5})/2$. Then, as the next exercises will help you to prove, the Fibonacci numbers are given by the formula known as the **Binet formula**:

$$F_n = (\phi^n - \bar{\phi}^n)/(\phi - \bar{\phi}). \tag{3}$$

Notice that the denominator in formula (3) is simply $\sqrt{5}$.

EXERCISE 10.2.   Verify that the Binet formula (3) gives F_0, F_1, F_2, and F_3 correctly, for the respective values of n.

EXERCISE 10.3.   Prove that

$$\phi^{n+1} - \bar{\phi}^{n+1} = (\phi^n - \bar{\phi}^n) + (\phi^{n-1} - \bar{\phi}^{n-1}).$$

EXERCISE 10.4.   Divide each term in the equation of Exercise 10.3 by $(\phi - \bar{\phi})$. Use the resulting generating formula together with the result of Exercise 10.2 to prove the Binet formula (3).

10.2   THE GOLDEN SECTION

The number $\phi = (1 + \sqrt{5})/2$ introduced above appears in connection with several geometry problems.

EXERCISE 10.5.   Divide a given line segment AB at a point C such that AB is in the same ratio to the part AC as AC is to the other part CB; that is, $AB/AC = AC/CB$. Show that the common ratio is ϕ. Such a division of a segment is called a **Golden Section**.

EXERCISE 10.6.   Let a rectangle have length l and width w. Remove from it a square of side w so as to leave a small rectangle similar to the original one (Fig. 10.2). Prove that the common ratio length/width $= \phi$.

EXERCISE 10.7.   Prove that the Golden Section is constructible by ruler and compass: Given AB, construct BD perpendicular to AB and half as long. Draw AD. Locate E on AD so that $ED = BD = AB/2$. Locate C on AB so that $AC = AE$. Prove that $AB/AC = AC/CB = \phi$. [See Figure 10.3.]

EXERCISE 10.8.   Cut AB in the Golden Section by a construction like that in Exercise 10.7, so that $AB/AC = \phi$. Bisect CB, labeling the midpoint M (Fig. 10.4). Draw the perpendicular to AB at M and locate T on it so that $AC = CT$. Draw AT and BT. Let $AC = x$, $CB = y$. Then from the choice of C, $(x+y)/x = x/y$. Show that $AT = x + y$. Show that the isosceles triangles ABT and TCB are similar. Then show that $\angle B = \angle ATB = \angle ATC + \angle CTB = \angle A + \angle A$, so that the angles of triangle ABT are $\angle A$, $2\angle A$, $2\angle A$, totaling $5\angle A = 180°$, so that $\angle A = 36°$. Prove that it is possible to construct a regular pentagon by ruler-and-compass construction.

10.3   DIVISIBILITY

The Fibonacci numbers have some curious divisibility properties.

Theorem 10.1   *No two consecutive Fibonacci numbers F_n, F_{n+1} have a prime factor p in common.*

Proof: See Exercise 10.9.

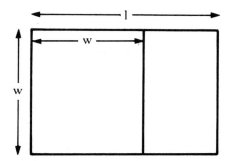

Figure 10.2   Golden section of a rectangle.

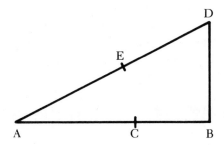

Figure 10.3   Ruler and compass construction of Golden Section.

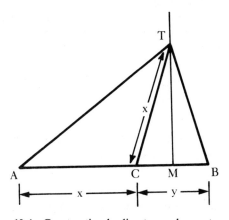

Figure 10.4   Construction leading to regular pentagon.

EXERCISE 10.9. Prove Theorem 10.1 by supposing that $p \mid F_n$ and $p \mid F_{n+1}$, deducing from the formula $F_{n+1} = F_n + F_{n-1}$ that $p \mid F_{n-1}$, and hence in turn $p \mid F_{n-2}, F_{n-3}, \ldots, F_2 = 1.$ ∎

EXERCISE 10.10. Let $\phi = (1+\sqrt{5})/2$ and $\bar{\phi} = (1-\sqrt{5})/2$, as in the Binet formula (3). Prove that $\phi^2 = \phi + 1$ and $\bar{\phi}^2 = \bar{\phi} + 1$, so that ϕ and $\bar{\phi}$ are the two roots of the equation $x^2 = x + 1$. Show that $\phi + \bar{\phi} = 1$ and $\phi\bar{\phi} = -1$.

In Exercise 10.1 we introduced the Lucas numbers, the generalized Fibonacci numbers with $u_1 = 1$ and $u_2 = 3$ such that $L_1 = 1$, $L_2 = 3$, $L_3 = 4, \ldots, L_{n+1} = L_n + L_{n-1}$. In the following lemma we establish a formula in ϕ and $\bar{\phi}$ for L_n, corresponding to the Binet formula for F_n.

Lemma 10.1   *The Lucas numbers L_n have the formula*

$$L_n = \phi^n + \bar{\phi}^n.$$

Proof: First, $L_1 = \phi^1 + \bar{\phi}^1 = ((1+\sqrt{5})/2) + ((1-\sqrt{5})/2) = 1.$ Also, $L_2 = \phi^2 + \bar{\phi}^2 = (\phi+1) + (\bar{\phi}+1) = (\phi+\bar{\phi}) + 2 = 1 + 2 = 3$ (see Exercise 10.10 for the identities used here). For larger n we rely on the generating relation $L_{n+1} = L_n + L_{n-1}$. Does the formula $\phi^n + \bar{\phi}^n$ have the same generating relation?

$$[\phi^n + \bar{\phi}^n] + [\phi^{n-1} + \bar{\phi}^{n-1}] = \phi^{n-1}(\phi+1) + \bar{\phi}^{n-1}(\bar{\phi}+1).$$

But according to Exercise 10.10, $\phi+1 = \phi^2$ and $\bar{\phi}+1 = \bar{\phi}^2$, so that $[\phi^n + \bar{\phi}^n] + [\phi^{n-1} + \bar{\phi}^{n-1}] = \phi^{n+1} + \bar{\phi}^{n+1}$. This, together with the verification for L_1 and L_2, shows that the formula represents the appropriate Lucas number for each n. ∎

In the next proof we shall need the fact that $(\phi^n + \bar{\phi}^n)$ always represents an integer. Since, as we now know, it represents a Lucas number, it does represent an integer.

EXERCISE 10.11. Use the generating relation $L_{n+1} = L_n + L_{n-1}$ to prove that each Lucas number is an integer.

EXERCISE 10.12. Are all Fibonacci numbers integers? Describe a proof paralleling Exercise 10.11.

Theorem 10.2   *Let $d \mid n$, where d and n are positive integers. Then $F_d \mid F_n$.*

Proof: From Definition 2.1 of divisibility, there is a quotient q for which $n = dq$. Use the Binet formula to write

$$F_n = F_{dq} = \frac{\phi^{dq} - \bar{\phi}^{dq}}{\phi - \bar{\phi}}$$

$$= \frac{\phi^d - \bar{\phi}^d}{\phi - \bar{\phi}}(\phi^{dq-d} + \phi^{dq-2d}\bar{\phi}^d + \cdots + \phi^d \bar{\phi}^{dq-2d} + \bar{\phi}^{dq-d}),$$

which is a multiple of $F_d = (\phi^d - \bar{\phi}^d)/(\phi - \bar{\phi})$. To be sure of the desired divisibility we have to prove that the multiplier of F_d is an integer. We pair the first and last terms, the second and the next-to-last, and so on, and apply Lemma 10.1.

$$(\phi^{dq-d} + \bar{\phi}^{dq-d}) + (\phi\bar{\phi})^d(\phi^{dq-3d} + \bar{\phi}^{dq-3d}) + \cdots$$

$$+ \begin{cases} (\phi\bar{\phi})^{rd-d}(\phi^d + \bar{\phi}^d) & \text{if} \quad q = 2r \\ (\phi\bar{\phi})^{rd} & \text{if} \quad q = 2r+1 \end{cases}$$

$$= L_{dq-d} + (-1)^d L_{dq-3d} + \cdots + \begin{cases} (-1)^{rd-d} L_d & \text{if} \quad q = 2r \\ (-1)^{rd} & \text{if} \quad q = 2r+1. \end{cases}$$

Then the multiplier of F_d is an integer, because it is a sum of Lucas numbers and 1 and their negatives. Then $F_d | F_n$. ∎

The next lemma and sequence of exercises culminate in a converse to Theorem 10.2, that if $F_m | F_n$, then, then $m | n$.

Lemma 10.2   *For n a positive integer, $2 | F_n$ if and only if $3 | n$.*

Proof: Let $3 | n$. Then by Theorem 10.2, $2 = F_3$ divides F_n. Conversely, suppose $2 | F_n$. Then $n > 1$, since $F_1 = 1$. By Theorem 10.1, $2 \nmid F_{n-1}$ and $2 \nmid F_{n+1}$. By Theorem 10.2, $3 \nmid n-1$ and $3 \nmid n+1$. Then $3 | n$. ∎

In the following exercises use the formulas

$$F_n = (\phi^n - \bar{\phi}^n)/\sqrt{5} \quad \text{and} \quad L_n = \phi^n + \bar{\phi}^n.$$

EXERCISE 10.13.   Show that $L_n = F_{2n}/F_n$ and that $L_n^2 - 5F_n^2 = 4(-1)^n$.

EXERCISE 10.14.   Use the last identity in Exercise 10.13 to prove that $(F_n, L_n) = 1$ or 2.

EXERCISE 10.15.   Prove that $2F_{m+n} = F_m L_n + F_n L_m$.

EXERCISE 10.16.   Use Exercise 10.14 and the identity proved in Exercise 10.15 to show that if F_m is odd, then $(F_m, F_{m+n}) | F_n$.

EXERCISE 10.17.   Suppose that $2|(F_m, F_{m+n})$. Then by Lemma 10.2, $2|F_m$ implies that $3|m$, and $2|F_{m+n}$ implies that $3|(m+n)$, so $3|n$. Since $L_{n+1} = L_n + L_{n-1}$, we can show, as in Theorem 10.1 for F_n, that no prime divides two consecutive Lucas numbers L_n. $L_3 = 4$ is divisible by 2. Suppose that $2|L_{3i}$. Then L_{3i+1} must be odd, so that $L_{3i+2} = L_{3i+1} + L_{3i}$ is odd. Then $L_{3i+3} = L_{3(i+1)} = L_{3i+2} + L_{3i+1}$ is the sum of two odd numbers, hence is even. This proves that since $3|n$, $2|L_n$.

Now use the identity of Exercise 10.15 in the form $2(F_{m+n} - F_m L_n/2) = F_n L_m$ to show that if $2|(F_m, F_{m+n})$, then, since $(F_m, L_m)|2$ as shown in Exercise 10.14, we again have $(F_m, F_{m+n})|F_n$, as in Exercise 10.16 for the case F_m odd.

Theorem 10.3   *Let a and b be positive integers and let $F_b|F_a$. Then $b|a$* [converse of Theorem 10.2].

Proof: Let $a = bq + r$, where $0 \leqslant r < b$, as in the Division Algorithm, Theorem 2.2. Then $F_a = F_{bq+r}$. From Theorem 10.2, $F_b|F_{bq}$ and by hypothesis $F_b|F_{bq+r}$. Then $F_b|(F_{bq}, F_{bq+r})|F_r$. This last division is justified by Exercise 10.16 if F_b is odd or by Exercise 10.17 if F_b is even. But $r < b$, so $r = 0$ and $b|a$. ∎

10.4   COMPLETENESS

Definition 10.1   *A sequence of positive integers $i_1, i_2, i_3, \ldots$ is* **complete** *with respect to the positive integers if every positive integer n has a* **representation** *as a sum of a finite number of integers from the sequence, none used more than once:*

$$n = i_{n_1} + i_{n_2} + \cdots + i_{n_m}, \quad \text{no two } i\text{'s the same.}$$

EXERCISE 10.18.   Check to see whether all requirements of Definition 10.1 are met for the sequence of powers of 2, $2^0, 2^1, 2^2, \ldots$ to be complete with respect to the positive integers.

We are going to prove that the Fibonacci sequence F_n is complete with respect to the positive integers, but first we need an identity connecting the nth Fibonacci number and the sum of the first $n-2$ Fibonacci numbers:

Lemma 10.3   *For $n > 2$,*

$$F_n = F_1 + F_2 + \cdots + F_{n-2} + 1 .$$

Proof: Consult the table in Figure 10.1. The number of baby pairs of rabbits after n months is F_n. For $n > 2$, the baby rabbits at the end of n months include one pair for each adult pair at the end of $n - 1$ months. But these adult pairs are all the baby pairs produced in previous months (that is, through $n - 2$) and mature at $n - 1$ months, plus the original breeding pair present when the experiment began; that is, $F_n = F_1 + F_2 + \ldots + F_{n-2} + 1$. ∎

Theorem 10.4   *The Fibonacci sequence F_n is complete with respect to the positive integers.*

Proof: We verify that the first few positive integers can be expressed as sums of Fibonacci numbers, each used only once:

$$1 = F_1, \; 2 = F_3, \; 3 = F_4, \; 4 = F_1 + F_4$$

(note that $F_3 + F_3$ is not the kind of representation we require, since it uses F_3 twice). Continuing,

$$5 = F_3 + F_4, \; 6 = F_1 + F_3 + F_4, \; 7 = F_1 + F_2 + F_3 + F_4.$$

We have used only the Fibonacci numbers through F_4 to express all the integers less than $F_6 = 8$. This will enable us to satisfy the finiteness requirement of Definition 10.1. Our proof is by mathematical induction. The induction hypothesis must be stated with care, because we want the provision $n > 2$, we want to include all the positive integers less than F_n, and we want to limit the number of Fibonacci numbers used in the representation.

Induction Hypothesis: Let n be an integer > 2. Each of the integers

$$1, 2, 3, \ldots, F_n - 1$$

has a representation as a sum of numbers in the set

$$\{F_1, F_2, F_3, \ldots, F_{n-2}\}$$

with no repetitions.

EXERCISE 10.19.   Verify that the preceding statement holds for $n = 3$, for $n = 4$, for $n = 5$, and for $n = 6$. Show that this establishes a basis for the induction.

Now we use the induction hypothesis together with the identity proved in Lemma 10.3 to show the result for $n + 1$. Write the representations we have assumed to exist for $1, 2, 3, \ldots, F_n - 1$, and add F_{n-1} to each one. This gives us representations for

$$1 + F_{n-1}, \; 2 + F_{n-1}, \; 3 + F_{n-1}, \; \ldots, \; \text{and} \; F_n - 1 + F_{n-1}$$

in terms of the set $\{F_1, F_2, F_3, ..., F_{n-2}, F_{n-1}\}$. Since $F_n - F_{n-1} = F_{n-2} \geqslant 1$ for $n > 2$, we have

$$F_n - 1 \geqslant F_{n-1}.$$

Then between the integers $1, 2, 3, ..., F_n - 1$ in the hypothesis and the new list $1 + F_{n-1}, 2 + F_{n-1}, ..., F_n - 1 + F_{n-1}$ there is no integer left without a representation. Then, since $F_n + F_{n-1} = F_{n+1}$, we have representations for the integers

$$1, 2, ..., F_{n+1} - 1$$

as sums of numbers in the set

$$\{F_1, F_2, F_3, ..., F_{n-1}\},$$

without repetitions. ∎

The completeness of the Fibonacci numbers can be applied to some of the ancient puzzles about weights, for with a set of weights of 1 lb., 2 lb., 3 lb., 5 lb., 8 lb., 13 lb., ..., and F_{n-2} lb., one can balance any integral weight less than F_n lb., as Theorem 10.4 shows. It can even be shown that if any one of the weights is lost from the set one can still balance any integral weight less than F_n lb. [The proof is based on the fact that we did not use F_{n-1} in the representation in Theorem 10.4.] However, if two weights are missing, then there are integral weights less than F_n that cannot be balanced by combinations of the remaining weights. Many puzzles are made more complex and interesting by allowing weights to be placed in either pan of the scale balance, which is equivalent to allowing negatives in the representation.

10.5   APPROXIMATION USING FIBONACCI NUMBERS

We see in
$$2 \cdot 5 - 3^2 = 1$$
$$5 \cdot 13 - 8^2 = 1$$
$$13 \cdot 34 - 21^2 = 1$$

some special cases of the following identity:

Theorem 10.5   *For n a positive integer,* $F_{2n-1} \cdot F_{2n+1} - F_{2n}^2 = 1.$

Proof: Using the Binet formula, we can write

$F_{2n-1} \cdot F_{2n+1} - F_{2n}^2$

$= [(\phi^{2n-1} - \bar{\phi}^{2n-1})/\sqrt{5}](\phi^{2n+1} - \bar{\phi}^{2n+1})/\sqrt{5} - [(\phi^{2n} - \bar{\phi}^{2n})/\sqrt{5}]^2$

$= (\phi^{4n} - \phi^{2n-1}\bar{\phi}^{2n+1} - \phi^{2n+1}\bar{\phi}^{2n-1} + \bar{\phi}^{4n})/5 - (\phi^{4n} - 2\phi^{2n}\bar{\phi}^{2n} + \bar{\phi}^{4n})/5$

$= -(\phi\bar{\phi})^{2n-1}[(\bar{\phi}^2 + \phi^2)/5 + 2\phi^2\bar{\phi}^2/5] = (\bar{\phi} + 1 + \phi + 1)/5 + 2/5$

$= 1.$ ∎

Lewis Carroll ("Alice in Wonderland") was a pen-name for the mathematician Charles Dodgson. Dodgson especially liked puzzles, and one of his favorites was a geometric "paradox" based on the identity of Theorem 10.5. Start with a paper square of side F_{2n} and separate it into four parts as in Figure 10.5.

It seems that the four parts of the square can be rearranged in a rectangle as shown, with an increase of one unit in area. The trick is that the sides do not quite meet along the diagonal of the rectangle, but enclose a slim parallelogram—the "magically" added unit of area.

EXERCISE 10.20.   Draw an 8×8 square on graph paper $(n = 4)$, separate it as in Figure 10.5, and try to convince someone that you can rearrange the four sections so as to increase the total area.

EXERCISE 10.21.   Prove that $F_{2n} \cdot F_{2n+2} - F_{2n+1}^2 = -1$ for n a positive integer.

We can rewrite the identities of Theorem 10.5 and Exercise 10.21 to find

$$\frac{F_{2n+1}}{F_{2n}} - \frac{F_{2n}}{F_{2n-1}} = \frac{1}{F_{2n-1} \cdot F_{2n}} \quad \text{and} \quad \frac{F_{2n+2}}{F_{2n+1}} - \frac{F_{2n+1}}{F_{2n}} = \frac{-1}{F_{2n} \cdot F_{2n+1}}.$$

Since the right member of each equation can be made as small as we want by taking n large, it appears that the ratio of each Fibonacci number to the preceding one tends to stay near some fixed number for large n. It can be shown that the fixed "limit" is $\phi = (1 + \sqrt{5})/2$. Because the difference between successive ratios alternates between $+$ and $-$, as shown in the two identities, the ratios oscillate about the limit ϕ (See Exercise 10.22).

EXERCISE 10.22.   Compute 1/1, 2/1, 3/2, 5/3, 8/5, 13/8, 21/13, and 34/21. Look up $\sqrt{5}$ in a table and compute $\phi = (1 + \sqrt{5})/2$. Put a check beside the ratios you have computed that are greater than ϕ.

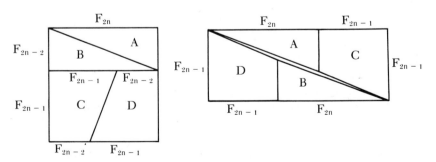

Figure 10.5   An apparent geometric paradox.

EXERCISE 10.23.   Describe how to use the Fibonacci numbers to get an approximation for $\sqrt{5}$, and use your method to get $\sqrt{5}$ correct to the nearest tenth.

Chapter 11 shows how to apply similar methods to approximate other numbers and gives us a link between Fibonacci numbers and Diophantine equations (Chapter 9).

In this book our attention is on "elementary number theory," which does not draw on limits and related concepts from calculus. However, by introducing approximation by an infinite sequence we show how methods from calculus can be applied as in "analytic number theory."

For more about Fibonacci numbers and related matters, look at issues of "The Fibonacci Quarterly," founded in 1963. Fibonacci numbers appear frequently in nature, notably in the numbers of clockwise and counter-clockwise spirals in the seed patterns of sunflowers. Perhaps you can suggest an explanation for this, based on the generating relation of the Fibonacci sequence. For illustrations of Fibonacci numbers in nature and simple explanations of some of the properties of the numbers, see "Fibonacci and Lucas Numbers" by Verner E. Hoggatt, Jr., Houghton Mifflin, 1969.

SUGGESTED PROJECTS

1. A talk for a Mathematics Club might begin with a description of Fibonacci's rabbit problem. The audience can help construct a table like that of Figure 10.1, and then a list of the first 20 Fibonacci numbers, say. Point out that no two consecutive numbers have a prime factor in common (Theorem 10.1). Observe that Theorems 10.2 and 10.3 hold, and put the two together to get: "One Fibonacci number divides another if and only if the index of the first divides the index of the second." Introduce $\phi = (1+\sqrt{5})/2$ and $\bar{\phi} = (1-\sqrt{5})/2$, and show how to use them to compute Fibonacci numbers. You might compute several to show that they match the ones generated one at a time by adding the previous two.

2. A demonstration of completeness can be made either separately, or in connection with Project 1. Prepare a set of cardboard "weights" labelled to represent 1 pound, 2 pounds, 3 pounds, 5 pounds, and so on, for Fibonacci numbers through, say, $F_8 = 21$, and show that you can combine them to balance any weight through $F_{10} = 55$. In fact, if you lose one of the set of weights, you can still balance any weight through 55.

3. Several experiments have been made showing the eye appeal of the Golden Section. Prepare pictures showing rectangles of comparable

area but different length-width ratios. Have as many people as possible say which shape they find most attractive. See whether most people choose a length-width ratio that is close to the Golden Section ϕ. You will get a better response if you describe the rectangles as representing possible shapes for a gift package or designs for a room or office building. A related experiment calls for people to judge where a line should be drawn across a rectangle for best appearance (Figure 10.2). This can be presented as a problem in artistic composition, such as where to place a division in a photograph, or where to place a ribbon tie on a gift package.

Cite examples of the Golden Section in man-made objects. Look for examples in nature.

4. Make a demonstration based on Exercise 10.20 on Lewis Carroll's paradox.

CHAPTER 11

Pell's Equation

In Chapter 10 we constructed Fibonacci numbers and Lucas numbers from $u_1 = 1$ and $u_2 = 1$ or 3, respectively, by using the recursive relation

$$u_{n+1} = u_n + u_{n-1}.$$

We can generate different sequences by changing the initial values, changing the recursive relation, or changing both. For instance, we might start with three initial terms, $u_1 = 1$, $u_2 = 2$, $u_3 = 3$, and use as a recursive relation $u_{n+2} = u_{n+1} + u_n u_{n-1}$ to generate later terms.

About 130 A.D. Theon of Smyrna gave a sequence of ratios for approximating $\sqrt{2}$, called "side-and-diagonal numbers":

$$d_1/s_1 = 1/1,\ 3/2,\ 7/5,\ 17/12,\ ...,\ d_n/s_n,\ ...$$

formed from two interrelated sequences. Each new denominator s_{n+1} is calculated from the previous numerator and denominator as

$$s_{n+1} = s_n + d_n, \tag{1}$$

and each numerator from the previous numerator and denominator as

$$d_{n+1} = 2s_n + d_n. \tag{2}$$

EXERCISE 11.1. Find the next two ratios in the sequence d_n/s_n. Compute the six ratios, look up $\sqrt{2}$, and put a check beside the ratios that are greater than $\sqrt{2}$. (Cf. Exercise 10.22.)

The side-and-diagonal number pairs s_n, d_n satisfy an identity similar to that of Theorem 10.5 and Exercise 10.21:

$$d_n^2 - 2s_n^2 = \pm 1, \tag{3}$$

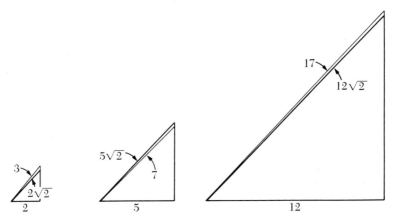

Figure 11.1   Side-and-diagonal numbers approximating $\sqrt{2}$.

with the sign of the difference alternating for successive pairs, as does the sign of $F_n \cdot F_{n+2} - F_{n+1}^2$ for Fibonacci numbers.

EXERCISE 11.2.   Verify that $d_n^2 - 2s_n^2$ oscillates between -1 and $+1$ for $n = 1, 2, 3,$ and 4. Prove from the generating relations that $d_{n+1}^2 - 2s_{n+1}^2 = -(d_n^2 - 2s_n^2)$. Argue that this substantiates the statement that $d_n^2 - 2s_n^2$ alternates between -1 and $+1$ as n increases.

If we divide equation (3) by s_n^2, getting

$$(d_n/s_n)^2 - 2 = \pm 1/s_n^2,$$

and note that the right member becomes negligible for large n, we see that d_n/s_n is indeed an approximation for $\sqrt{2}$.

The following exercises develop sequences of side-and-diagonal numbers for estimating $\sqrt{3}$, analogous to the above sequences for $\sqrt{2}$.

EXERCISE 11.3.   Form the first 6 ratios $\bar{d}/\bar{s}$, from the sequences

$$\bar{d}_1 = 1, \bar{d}_2 = 4, ..., \bar{d}_{n+1} = 3\bar{s}_n + \bar{d}_n, ...$$
$$\bar{s}_1 = 1, \bar{s}_2 = 2, ..., \bar{s}_{n+1} = \bar{s}_n + \bar{d}_n, ...$$

EXERCISE 11.4.   Compute the six ratios of Exercise 11.3, look up $\sqrt{3}$, and put a check beside those ratios that exceed $\sqrt{3}$.

EXERCISE 11.5.   Show that $\bar{d}_1^2 - 3\bar{s}_1^2 = -2$ and that $\bar{d}_{n+1}^2 - 3\bar{s}_{n+1}^2 = -2(\bar{d}_n^2 - 3\bar{s}_n^2)$. Conclude that $\bar{d}_n^2 - 3\bar{s}_n^2 = (-2)^n$.

EXERCISE 11.6.   Divide the formula you derived in Exercise 11.5 by $\bar{s}_n^2$. How large an n do you need to guarantee that $(\bar{d}_n/\bar{s}_n)^2 - (\sqrt{3})^2$ is less than 1 in absolute value? Let f be a fraction in the range $-1 < f < 1$. What happens to the size of f^n as n increases?

More material on these sequence approximations can be found in "Side-and-Diagonal Numbers," by F. V. Waugh and Margaret W. Maxfield, *Mathematics Magazine*, Vol. 40, No. 2, March 1967, pp. 74–83.

The equation $x^2 - ry^2 = 1$, where r is an integer, was named for John Pell by Leonard Euler, because of a method of solution he mistakenly attributed to Pell. In fact, by now any equation of the form
$$x^2 - ry^2 = N,$$
where r and N are integers, is known as a Pell's equation.

EXERCISE 11.7.   Consider the side-and-diagonal numbers s_n, d_n of equation (3). Show that for n odd they satisfy one Pell's equation and for n even they satisfy another.

EXERCISE 11.8.   Let $(x, y) = (d, s)$ satisfy Pell's equation $x^2 - ry^2 = 1$. If $(d - s\sqrt{r})^n = f - e\sqrt{r}$, then $(x, y) = (f, e)$ satisfies the same equation. Prove this by supposing that $(d - s\sqrt{r})^n = f - e\sqrt{r}$ provides a solution and showing that $(d - s\sqrt{r})^{n+1} = (d - s\sqrt{r})(f - e\sqrt{r})$ yields a solution.

EXERCISE 11.9.   In Chapter 9 we investigated several special cases of Pell's equation. What values for r and N have we discussed? One or both of these integers may be taken to be negative. Notice the discussion on sums of two squares under the four-square problem.

SUGGESTED PROJECT

A report or talk can be based on side-and-diagonal numbers for approximating square roots. A figure like that of Figure 11.1 can illustrate the approximation. Observe that the successive approximations oscillate around the square root.

CHAPTER 12

Magic Numbers

12.1 PERFECT NUMBERS

Are you superstitious about the number 13? Is there a number you consider especially lucky for you? If so, then you have experienced a common belief that some positive integers have "magic" in them. Ancient Greeks thought that the clue to the mystical nature of a number n lay in its divisors. They compared n with the sum $\sigma^-(n)$ of the divisors of n that are less than n.

If $\sigma^-(n) < n$, they said n was "deficient".

If $\sigma^-(n) > n$, they said n was "abundant".

If $\sigma^-(n) = n$, they said n was "*perfect*", and considered such an n a "good" number.

The divisors of 12 less than 12 are 1, 2, 3, 4, and 6.

$$\sigma^-(12) = 1 + 2 + 3 + 4 + 6 = 16 > 12,$$

so 12 is not a perfect number.

Figure 12.1 Test for a perfect number.

EXERCISE 12.1. Following Figure 12.1, show that 4 and 8 are not perfect, but that 6 is perfect.

EXERCISE 12.2. Prove that no prime p is perfect.

EXERCISE 12.3. Find a perfect number between 25 and 30.

It will simplify notation and our results generally if we replace the Greeks' $\sigma^-(n)$ with $\sigma(n)$, the sum of *all* the positive divisors of n, including n itself. We also define $\tau(n)$, which tells us how many divisors n has.

Definition 12.1   *Let n be a positive integer. Then*

$$\sigma(n) = \sum_{d|n} d \quad and \quad \tau(n) = \sum_{d|n} 1 \ .$$

This notation means that $\sigma(n)$ is the sum of all positive integers d such that d divides n, and $\tau(n)$ is the number of such d's.

$$\sigma(28) = \sum_{1,2,4,7,14,28} d = 1 + 2 + 4 + 7 + 14 + 28 = 56$$

$$\tau(28) = \sum_{d|28} 1 = 1 + 1 + 1 + 1 + 1 + 1 = 6$$

Figure 12.2   Computation of σ and τ.

EXERCISE 12.4.   Following Figure 12.2, find $\sigma(6)$ and $\tau(6)$. Find $\sigma(12)$ and $\tau(12)$.

EXERCISE 12.5.   Find $\sigma(n)$ and $\tau(n)$ for $n = 25$, $n = 26$, $n = 27$, and $n = 30$.

EXERCISE 12.6.   Let p be a prime. Find $\sigma(p)$ and $\tau(p)$.

EXERCISE 12.7.   Give a formula for finding $\sigma(n)$ if you have $\sigma^-(n)$, the sum of the divisors less than n.

EXERCISE 12.8.   Referring to Exercise 12.7, justify the following definition:

Definition 12.2   *A* **perfect number** *is a positive integer n for which $\sigma(n) = 2n$.*

EXERCISE 12.9.   Find $\sigma(3)$, $\sigma(3^2)$, $\sigma(3^3)$, and $\sigma(3^4)$. Also find $\tau(n)$ for each of these powers of 3.

EXERCISE 12.10.   Find $\sigma(2)$, $\sigma(2^2)$, $\sigma(2^3)$, $\sigma(2^4)$, and $\sigma(2^5)$. Also find $\tau(n)$ for these five powers of 2.

EXERCISE 12.11.   Find $\sigma(n)$ and $\tau(n)$ for $n = 5$, for $n = 7$, and for $n = 35$.

EXERCISE 12.12.   Find $\sigma(n)$ and $\tau(n)$ for $n = 11, n = 13$, and for $n = 143$.

As we can observe from Theorem 7.3, the Euler ϕ-function, $\phi(n)$, which counts the positive integers less than or equal to n that are prime to n, is a **multiplicative function:** By this we mean that if m and n are relatively prime, then $\phi(mn) = \phi(m)\phi(n)$. We shall show that $\sigma(n)$ and $\tau(n)$ also have this property, so that if $(m, n) = 1$, we can find the function of their product as the product of the functions.

Theorem 12.1   *Let m and n be positive integers, with $(m, n) = 1$. Then*

$$\sigma(mn) = \sigma(m)\sigma(n)$$

and

$$\tau(mn) = \tau(m)\tau(n) ;$$

that is, σ and τ are multiplicative functions.

Proof: Follow the special case in Figure 12.3.

Let $m = 4$ and $n = 7$. $m = 4$ has divisors $1, 2, 4$, and $n = 7$ has divisors $1, 7$. The product $mn = 28$ has divisors

$$1 \cdot 1, 2 \cdot 1, 4 \cdot 1$$

$$1 \cdot 7, 2 \cdot 7, 4 \cdot 7$$

Figure 12.3   Divisors of 28.

Let the divisors of m be $m_1 = 1, m_2, m_3, \ldots, m_s = m$. Let the divisors of n be $n_1 = 1, n_2, n_3, \ldots, n_t = n$. Every product $m_i n_j$ for some i from 1 to s and for some j from 1 to t is a divisor of mn, because $m_i \mid m$ and $n_j \mid n$.

We want to show conversely that every divisor d of mn appears as one of the products $m_i n_j$. Any divisor d of mn can be written uniquely as a product of prime powers p^w (Theorem 4.1). Each prime power p^w divides m or n but not both, since m and n are relatively prime (Corollary 4.2). The product of the various prime power factors of d that divide m is a divisor m_i of m, and the product of the other prime power factors of d is a divisor n_j of n.

Then the divisors of mn are

$$m_1 n_1, m_2 n_1, \ldots, m_s n_1,$$

$$m_1 n_2, m_2 n_2, \ldots, m_s n_2,$$

$$\vdots$$

$$m_1 n_t, m_2 n_t, \ldots, m_s n_t.$$

There are t rows, each containing s divisors; therefore, there are st divisors in all. Then $\tau(mn) = st = \tau(m)\tau(n)$.

The sum of the first row of divisors is $(\sum_{i=1}^{s} m_i)n_1 = [\sigma(m)]n_1$. The sum of the second row is $(\sum_{i=1}^{s} m_i)n_2 = [\sigma(m)]n_2$, and so on, so that the sum of all st divisors is $[\sigma(m)](\sum_{j=1}^{t} n_j) = \sigma(m)\sigma(n)$. Then $\sigma(mn) = \sigma(m)\sigma(n)$. ∎

EXAMPLE. From the prime-power factorization (Theorem 4.1), $180 = 2^2 \cdot 3^2 \cdot 5$, we can compute $\sigma(180)$ as the product of $\sigma(2^2)$, $\sigma(3^2)$, and $\sigma(5)$, by Theorem 12.1. Also, by Theorem 12.1, $\tau(180) = \tau(2^2)\tau(3^2)\tau(5)$.

EXERCISE 12.13. Referring to previous exercises and illustrations, show that $\sigma(mn) = \sigma(m)\sigma(n)$ and $\tau(mn) = \tau(m)\tau(n)$ for the cases

$$m = 4, n = 7$$
$$m = 4, n = 3$$
$$m = 5, n = 7$$
$$m = 11, n = 13.$$

By Theorem 4.1, any positive integer can be factored uniquely into prime powers, and by Theorem 12.1 we can find the functions σ and τ for the integer if we can find them for the prime-power factors. In the next theorem we show how to evaluate the functions for a power of a prime.

Theorem 12.2 *Let p and w be positive integers, p a prime. Then*

$$\sigma(p^w) = \frac{p^{w+1}-1}{p-1}$$

and $\tau(p^w) = w+1$.

Proof: All the divisors of p^w are powers of p, if we use the convention $p^0 = 1$. They are

$$1, p, p^2, \ldots, \text{and } p^w.$$

Then

$$\sigma(p^w) = 1 + p + p^2 + \cdots + p^w = \frac{p^{w+1}-1}{p-1}$$

and $\tau(p^w) = w+1$, the number of the divisors. ∎

EXERCISE 12.14. Use Theorem 4.1, Theorem 12.2, and Theorem 12.1 to find $\sigma(n)$ and $\tau(n)$ for $n = 496$. Then show that 496 satisfies Definition 12.2 of a perfect number.

EXERCISE 12.15. You have found the smallest three perfect numbers, 6, 28, and 496. Show that each of them has the form $2^{w-1}(2^w - 1)$, where $2^w - 1$ is a prime.

In Exercise 12.15 you have noticed a form that turns out to be the form for all even perfect numbers.

Theorem 12.3 *Let $2n$ be an even positive integer. Then $2n$ is a perfect number if and only if there is a positive integer w for which $2n = 2^{w-1}(2^w - 1)$, with $2^w - 1$ a prime.*

Proof: First, prove the "if" part:

EXERCISE 12.16.   Suppose that $2^w - 1$ is a prime. Then from Theorem 12.2, $\sigma(2^w - 1) = $ _____? $\sigma(2^{w-1}) = $ _____? Since $2^w - 1$ is odd, we have $(2^{w-1}, 2^w - 1) = 1$. Then by Theorem 12.1, $\sigma(2^{w-1}, (2^w - 1)) = $ _____? $2[2^{w-1}(2^w - 1)] = $ _____? Then by Definition 12.2, $2^{w-1}(2^w - 1)$ is a perfect number.

Next, prove the "only if" part. Suppose $2n$ is an even perfect number. We want to prove that $2n$ is of the form $2^{w-1}(2^w - 1)$ with $2^w - 1$ a prime. Let y be the highest power of 2 that divides $2n$, so that $2n = 2^y m$, and $2 \nmid m$. Since $2n$ is even, we have $y \geqslant 1$. Since $2n$ is assumed perfect, we have

$$\sigma(2^y m) = 2(2^y m)$$

or

$$\frac{2^{y+1} - 1}{2 - 1} \cdot \sigma(m) = 2^{y+1} m.$$

The factor $2^{y+1} - 1$ is odd, so it cannot divide 2^{y+1} and thus must divide m. We can write $m = (2^{y+1} - 1)k$, then, for some positive integer k. Using this, we have

$$(2^{y+1} - 1)\sigma(m) = 2^{y+1}(2^{y+1} - 1)k,$$

or

$$\sigma(m) = 2^{y+1}k = (2^{y+1} - 1 + 1)k = (2^{y+1} - 1)k + k = m + k.$$

Since $\sigma(m)$ stands for the sum of the positive divisors of m, and m and k are both positive divisors of m, these two must constitute all those divisors, one of which must be 1. Thus $k = 1$ and $m = 2^{y+1} - 1$, a prime. If we let w stand for $y + 1$, we find the required form $2n = 2^{w-1}(2^w - 1)$, with $2^w - 1$ a prime. ∎

EXERCISE 12.17.   Find the fourth even perfect number in order of increasing size.

Since the criterion for even perfect numbers hinges on whether $2^w - 1$ is a prime, we are led to look at numbers of that form.

Theorem 12.4   *Let w be a positive integer. If $2^w - 1$ is a prime, then w is a prime.*

Proof: It is easy to prove this theorem in its contrapositive form: "If w is not a prime, then $2^w - 1$ is not a prime." If w is not a prime it is a unit or it is composite. If $w = 1$, then $2^w - 1 = 1$ and so is not a prime. Then suppose $w = xy$, a composite integer. Then $2^w - 1$ can be separated into factors:

$$2^w - 1 = 2^{xy} - 1 = (2^x - 1)(2^{xy-x} + 2^{xy-2x} + \ldots + 1). \quad \blacksquare$$

EXERCISE 12.18.   Factor $2^{15} - 1$ after the manner of the above proof.

From Theorem 12.4 we see that we need consider only forms $2^p - 1$ where p is a prime, in our search for primes $2^p - 1$ that yield even perfect numbers. Numbers of the form $2^p - 1$ are called **Mersenne numbers,** and those of them that happen to be primes are called **Mersenne primes.** Marin Mersenne (1588–1648) was a Franciscan friar, who conducted his large scientific correspondence from various Parisian monasteries during most of his life. He kept his contemporaries, Descartes, Fermat, and others, busy with challenging conjectures and problems, always urging them to work on perfect numbers. To this day we do not know whether there are infinitely many Mersenne primes, and so we do not know whether there are infinitely many even perfect numbers.

At least we know of *some* even perfect numbers. As for odd perfect numbers, we are not so well off. Not one has ever been found and it is known that if there is one it must be bigger than one trillion. On the other hand, no one has been able to prove that there cannot be an odd perfect number.

In Exercise 12.2 you proved that no prime p is a perfect number. There are several other special cases of odd numbers that you can decide in the negative.

EXERCISE 12.19.   Let p^w be a power of an odd prime. Show that p^w cannot be a perfect number by using the inequality

$$\sigma(p^w) = p^w + p^{w-1} + p^{w-2} + \ldots + 1 = p^w + \frac{p^w - 1}{p - 1} < p^w + p^w.$$

EXERCISE 12.20.   Let p and q be distinct odd primes. Expand the inequality $(p-1)(q-1) > 2$ to show that $pq > p + q + 1$. Then show that $\sigma(pq) < 2pq$ and conclude that pq is not a perfect number.

It is known that if n is an odd perfect number, then all the prime powers in its factorization must be even powers except for one. This one must have the form p^w, where both p and w are congruent to $1 \pmod 4$, that is, leave remainders of 1 upon division by 4. (See *Excursions Into Mathematics* by Anatole Beck et al., Worth Publishers, 1969, page 128 and following.)

Donald Gillies found the largest known Mersenne prime in 1963, for $p = 11213$. Thus the largest known even perfect number is the twenty-third one,

$$2^{11212} M_{11213} = 2^{11212}(2^{11213} - 1),$$

which, expanded, is a number of 6,751 digits.

EXERCISE 12.21.   As we have seen, it is not known whether there are infinitely many perfect numbers. However, you can show that there are infinitely many "deficient" numbers, that is, numbers n for which $\sigma(n) - n < n$, by considering $n = p$, a prime, and using Euclid's theorem about the infinitude of primes.

EXERCISE 12.22.   Show that there are infinitely many "abundant" numbers, that is, numbers n for which $\sigma(n) - n > n$, by considering $n = 2^w 3, w > 1$.

12.2 AMICABLE NUMBERS

Ancient Greeks called two positive integers m and n **amicable** (or "friendly"), if the sum of the divisors of m less than m equals n and the sum of the divisors of n less than n equals m.

EXERCISE 12.23.   Using the notation of Definition 12.1, show that m and n are amicable if

$$\sigma(m) = \sigma(n) = m + n.$$

Pythagoras found the smallest pair of amicable numbers, 220 and 284.

EXERCISE 12.24.   Show that 220 and 284 are amicable.

EXERCISE 12.25.   For what numbers n do n and n [that is, $m = n$ and n] form an amicable pair?

Fermat, Descartes, and Euler found pairs of amicable numbers, as have several contemporary mathematicians. We do not know whether there are infinitely many pairs nor whether there can be amicable pairs having one even and one odd number. From the example $m = 3^3 \cdot 5 \cdot 7 \cdot 11$ and $n = 3 \cdot 5 \cdot 7 \cdot 139$, we know there are pairs both of which are odd. From Exercise 12.24, there are pairs both of which are even.

12.3 MAGIC SQUARES

A magic square of degree n is an n by n square arrangement of the positive integers $1, 2, 3, ..., n^2$ placed so that each row, each column, and each diagonal sums to a fixed number.

Figure 12.4 shows a magic square of degree 4.

4	10	15	5	row sums equal 34
7	13	12	2	$(7 + 13 + 12 + 2$, for instance$)$
14	8	1	11	column sums equal 34
9	3	6	16	$(15 + 12 + 1 + 6$, for instance$)$

diagonal sums equal 34

$(5 + 12 + 8 + 9 = 4 + 13 + 1 + 16)$

Figure 12.4   A magic square of degree 4.

EXERCISE 12.26.   Show that the common sum in a magic square of degree n is $n(n^2 + 1)/2$.

EXERCISE 12.27.   The magic square of degree 1 is simply $\boxed{1}$ Show by experiment that there can be no magic square of degree 2. Notice that

$$\begin{array}{|cc|} \hline 1 & 2 \\ 3 & 4 \\ \hline \end{array} \quad \text{and} \quad \begin{array}{|cc|} \hline 3 & 4 \\ 1 & 2 \\ \hline \end{array}$$

for instance, are essentially the same trial squares, since the row sums, column sums, and diagonal sums are the same collection in the two squares.

EXERCISE 12.28.   Complete this array so as to make it a magic square of degree 3:

8	1	
	5	7

Here is a method due to A. Mąkowski for building larger magic squares from smaller ones. Let S be a magic square of degree m and T a magic square of degree n. Replace the integer i in the square S by the square T with $n^2(i-1)$ added to each entry. The result is a magic square of degree mn.

EXERCISE 12.29.   Let S be the magic square

$$S = \begin{array}{ccc} 1 & 5 & 9 \\ 6 & 7 & 2 \\ 8 & 3 & 4 \end{array} \quad \text{and let } T = \begin{array}{cccc} 2 & 13 & 8 & 11 \\ 12 & 7 & 14 & 1 \\ 15 & 4 & 9 & 6 \\ 5 & 10 & 3 & 16 \end{array}$$

Then $m = 3$ and $n = 4$.

Form a magic square of degree 12 by Mąkowski's construction. The "5" in S will be replaced by the square T with $n^2(i-1) = 4^2(5-1) = 64$ added to each entry:

$$\begin{array}{cccc} 66 & 77 & 72 & 75 \\ 76 & 71 & 78 & 65 \\ 79 & 68 & 73 & 70 \\ 69 & 74 & 67 & 80 \end{array}$$

EXERCISE 12.30.   Prove by Mąkowski's construction that there exist magic squares of degree greater than any integer N.

Magic squares have been generalized in several different ways. Some squares include integers other than the first n^2 integers; some are restricted to consecutive integers, and some are not. The magic square in Figure 12.5 is made up entirely of primes.

83	29	101
89	71	53
41	113	59

Figure 12.5 A magic square of primes.

Many books on mathematical recreations, diversions, or puzzles include material on magic squares. Material for a report or talk can be found in an article "Magic Squares" by Richard Thiessen in "The Mathematical Log", reprinted in *More Chips from the Mathematical Log*, ed. Josephine P. Andree, published by Mu Alpha Theta, Univ. of Okla., Norman, Oklahoma.

SUGGESTED PROJECTS

1. A search for perfect numbers encourages practice in factoring integers, not as an end in itself, but as a way of finding all the divisors. This provides additional insight into the role of divisors and in a way that parallels the evolution of number theory. The search for perfect numbers was central to much of early number theory, and it can provide the same kind of incentive for us today.

2. Magic squares provide an interesting topic for even a very elementary audience. They can be used as a setting for practice in addition, more congenial than pages of exercises. For instance, check the examples of magic squares in this chapter to see whether they are indeed "magic." Construct some squares of your own, placing your own restrictions on the entries—all primes, famous dates, numbers from your address, birthday, and so on.

CHAPTER 13

Figures with Shapes

Many a hopeful play on the words "figure" and "shape" has been offered for comic relief in school mathematics. The ancients used to call the numbers 1, 3, 6, 10, 15, ... **triangular numbers,** because each counts a number of dots that can be arranged evenly in an equilateral triangle, as shown in Figure 13.1.

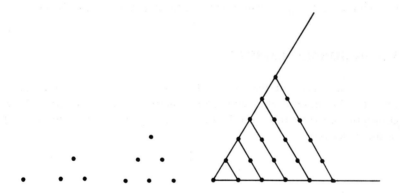

Figure 13.1 Triangular numbers.

If you analyze the formation of triangular numbers, you will see that each new one is formed from the one before by adding a row containing one more dot than the previously added row.

$$0 + 1 = 1$$
$$1 + 2 = 3$$
$$3 + 3 = 6, \text{ etc.}$$

Let t_n stand for the nth triangular number, with $t_1 = 1$, $t_2 = 3$, and so on. Then t_n is obtained from t_{n-1} by adding n: $t_n = t_{n-1} + n$, $n > 0$, where we let $t_0 = 0$. From this recursive definition we can find a formula for the kth triangular number if we sum t_n from $n = 1$ to $n = k$.

$$\sum_{n=1}^{k} t_n = \sum_{n=1}^{k} (t_{n-1} + n) = \sum_{n=1}^{k} t_{n-1} + \sum_{n=1}^{k} n$$

or

$$\sum_{n=1}^{k} t_n = \sum_{i=0}^{k-1} t_i + k(k+1)/2,$$

so that

$$t_k = \sum_{n=1}^{k} t_n - \sum_{n=0}^{k-1} t_n = k(k+1)/2.$$

EXERCISE 13.1.   Identify t_k as one of the binomial coefficients. Complete this diagram, showing the combinations of five circles taken two at a time:

Can you describe a connection between the diagram and the formula?

13.1 m-GONAL NUMBERS

Triangles are not the only polygons that have been connected to numbers. We have already considered **square numbers** in several different connections. Notice in Figure 13.2 how each square can be found from the next smaller one.

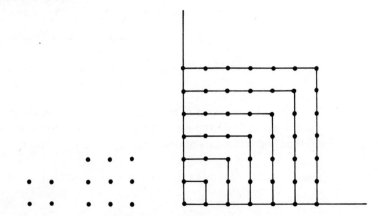

Figure 13.2   Square numbers.

$$0 + 1 = 1$$
$$1 + 3 = 4$$
$$4 + 5 = 9$$
$$9 + 7 = 16, \text{ etc.}$$

The added **gnomon** each time is the next odd number. If we let s_n stand for the nth square number with $s_1 = 1$, $s_2 = 4$, we have for $n > 0$ $s_n = s_{n-1} + 2n - 1$. We take $s_0 = 0$.

EXERCISE 13.2.   Write the familiar formula for s_n as the square of the nth positive integer. Using the formula, show that $s_n = s_{n-1} + 2n - 1$.

Figure 13.3 shows how **pentagonal**, or **5-gonal**, numbers are formed.

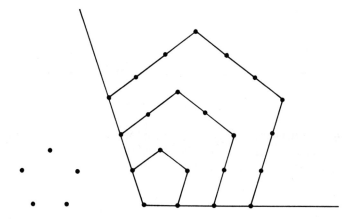

Figure 13.3   Pentagonal, or 5-gonal, numbers.

We have

$$0 + 1 = 1$$
$$1 + 4 = 5$$
$$5 + 7 = 12$$
$$12 + 10 = 22, \text{ etc.}$$

The added **gnomon** in this case is from the arithmetic progression $1 + 3k$: $1, 4, 7, 10, 13, \ldots$. We have for the nth pentagonal number p_5^n

$$p_5^n = p_5^{n-1} + 1 + 3(n-1).$$

Summing n from 1 to k, we can get a formula for p_5^k:

$$\sum_{n=1}^{k} p_5^n = \sum_{n=1}^{k} p_5^{n-1} + k + \frac{3k(k+1)}{2} - 3k,$$

so that

$$p_5^k = \sum_{n=1}^{k} p_5^n - \sum_{i=1}^{k-1} p_5^i = \frac{3k(k+1)}{2} - 2k = \frac{3k^2 - k}{2}.$$

EXERCISE 13.3.   Use the formula to find the first 7 pentagonal numbers.

Now we generalize from triangular or 3-gonal numbers, square or 4-gonal numbers, and pentagonal or 5-gonal numbers to m-gonal numbers. For $m = 6$ these are called hexagonal numbers, for $m = 7$ heptagonal, and so on. We take the 0th m-gonal number to be 0 for each m and take the first, p_m^1, to be 1 for each m. Then each new m-gonal number is formed from the preceding by adding the next gnomon from the arithmetic progression $1 + (m-2)k$.

$$p_m^n = p_m^{n-1} + 1 + (m-2)(n-1).$$

By summing on n we can get a general formula for the kth m-gonal number:

$$p_m^k = (2 + (m-2)(k-1))k/2.$$

Diophantus gave a more complicated formula equivalent to this.

As we shall see, the different m-gonal numbers are related to each other. Try summing pairs of consecutive triangular numbers.

$$p_3^2 + p_3^3 = 3 + 6 = 9 = p_4^3$$
$$p_3^3 + p_3^4 = 6 + 10 = 16 = p_4^4$$
$$p_3^6 + p_3^7 = 21 + 28 = 49 = p_4^7$$

EXERCISE 13.4.   Show from the formulas that

$$p_3^k + p_3^{k+1} = p_4^{k+1}.$$

EXERCISE 13.5.   Prove that the kth pentagonal number is the sum of the kth square and the $(k-1)$st triangular number. Verify this for the first 4 cases.

EXERCISE 13.6.   Show that the kth $(m+1)$-gonal number can be obtained from the kth m-gonal number by adding the $(k-1)$st triangular number: $p_{m+1}^k = p_m^k + p_3^{k-1}$. Show how this result implies the theorems of Exercise 13.4 and 13.5.

EXERCISE 13.7.   Fill in the blanks in this table showing p_m^k for $3 \leqslant m \leqslant 7$ and $1 \leqslant k \leqslant 10$.

		k									
		1	2	3	4	5	6	7	8	9	10
	3	1	3	6	10		21				
	4	1	4	9	16			49			
m	5	1	5	12	22				92		
	6	1	6	15	28					153	
	7	1	7	18	34						235

EXERCISE 13.8.   Verify the result of Exercise 13.6 in the table of Exercise 13.7.

EXERCISE 13.9.   Around the beginning of the second century A.D. Plutarch gave the theorem that if a triangular number is multiplied by 8, and then 1 is added, the result is a square. Prove this theorem.

EXERCISE 13.10.   Prove this theorem due to Bachet:

$$p_m^k = p_3^k + (m-3)p_3^{k-1}$$

Several mathematicians have treated the following problems: Which integers are *m*-gonal numbers? Which integers are both triangular and square? both square and pentagonal? Both *m*-gonal and *n*-gonal? How can we determine for which *m* values a given integer is an *m*-gonal number?

EXERCISE 13.11.   Show that 36 is a triangular number, a square number, a 13-gonal number, and a 36-gonal number.

Diophantus based the following test on the formula

$$p_m^k = (2 + (m-2)(k-1))k/2.$$

To find for which *m* values a given integer *b* is *m*-gonal, form 2*b* and express it as a product of two factors in all non-trivial ways (each factor > 1). Call the smaller factor *k*. Subtract 2 from the larger factor. If this difference is divisible by $k-1$, call the quotient $m-2$ and *b* is the *m*-gonal number p_m^k.

EXERCISE 13.12.   Use Diophantus' test on $b = 36$ and compare with Exercise 13.11.

Another question that has attracted interest is: Which positive integers can be written as the sum of two triangular numbers? or as the sum of three triangular numbers? or as the sum of three pentagonal numbers? With our convention $p_m^0 = 0$ for each m, we can write a generalization Fermat made of the four-square theorem (see Chapter 9) in a very simple way:

Every positive integer is expressible as the sum of m m-gonal numbers.

Fermat promised a whole book on this subject. In fact, he wrote: "I was the first to discover the very beautiful and entirely general theorem that every number is either triangular or the sum of 2 or 3 triangular numbers; every number is either a square or the sum of 2, 3 or 4 squares; either pentagonal or the sum of 2, 3, 4 or 5 pentagonal numbers; and so on ad infinitum, whether it is a question of hexagonal, heptagonal or any polygonal numbers. I cannot give the proof here, which depends upon numerous and abstruse mysteries of numbers; for I intend to devote an entire book to this subject and to effect in this part of arithmetic astonishing advances over the previous known limits." No such book ever appeared, and the first published proof of the complete generalization was by Cauchy in 1813–15.

13.2 PYRAMIDAL NUMBERS

The pyramidal numbers of various base shapes count the number of balls that can fill a pyramidal mold to a given level, and can be found by summing m-gonal numbers if the base shape is an m-gon.

The tetrahedral numbers are 1, $1+3 = 4$, $4+6 = 10$, $10+10 = 20$, and so on (see Figure 13.4).

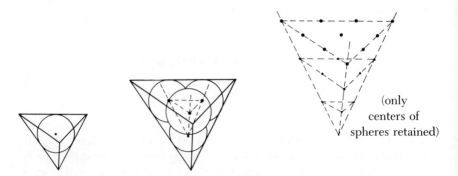

(only centers of spheres retained)

Figure 13.4   Tetrahedral numbers.

There is a formula for the kth pyramidal number p_m^k of m-gonal base shape in terms of the kth m-gonal number p_m^k:

$$p_m^k = \frac{k+1}{6}(2p_m^k + k).$$

13.3   PASCAL'S TRIANGLE

Pascal's triangle, shown in Figure 7.3, is yet another example of positive integers thought of in a geometric sense. See Exercise 13.1 for a connection with triangular numbers. Incidentally, Pascal called the triangle simply the "arithmetic triangle," and it was known by the end of the eleventh century, well before Pascal lived, among Arab mathematicians.

SUGGESTED PROJECT

Polygonal numbers lend themselves well to displays. Posters can show how triangular numbers (Figure 13.1), square numbers (Figure 13.2), and pentagonal numbers (Figure 13.3) can be generated. Calculating polygonal numbers by formula presents arithmetic practice in an interesting setting. Some of the relationships among polygonal numbers can be discovered through observation (Exercises 13.4 and 13.5). Use the formula of Exercise 13.6 to find entries for the table of Exercise 13.7.

Figure 13.4 suggests a physical model of tetrahedral numbers.

CHAPTER 14

Inside Integers

14.1 THE MÖBIUS FUNCTION

Mathematicians of antiquity tried to look "inside" an integer by studying its divisors (see Chapter 12). In this chapter we introduce the Möbius function, which can be used to x-ray certain kinds of integers, and in the process we discuss numerical functions in general.

A **numerical function** is a rule that assigns to each positive integer some value. For example:

the Euler phi-function, $\phi(n)$, assigns to each positive integer n the number of integers less than or equal to n that are prime to n.

the function $\sigma(n)$ assigns to each positive integer n the sum of its positive divisors.

the function $\tau(n)$ assigns to each positive integer n the number of its positive divisors.

These functions are called "numerical" to show that they are defined for positive integers. More generally, mathematical functions are defined over the real numbers, the unit interval, and geometrical figures, for example.

EXERCISE 14.1. Show that $f(n) = n$ is a numerical function. Find its value for $n = 1$, 10, and 20.

EXERCISE 14.2. Show that $f(n) = 1$ is a numerical function. Find its value for $n = 1$, 10, and 20.

EXERCISE 14.3. Show that $f(n) = n(n+1)/2$ is a numerical function. Find its value for $n = 1$, 10, and 20.

The functions $\phi(n)$, $\sigma(n)$, and $\tau(n)$ are **multiplicative** functions $f(n)$, which means that if m and n are relatively prime integers, then

$$f(mn) = f(m)f(n).$$

This property simplifies computing function values, for, by the Fundamental Theorem of Arithmetic (Theorem 4.1), each integer has a unique factorization into prime powers which are relatively prime to each other.

Now we introduce another numerical function. It, too, is multiplicative.

Definition 14.1 *The Möbius function $\mu(n)$ is defined for positive integers n by*

$$\mu(n) = \begin{cases} 1 & \textit{if } n = 1 \\ 0 & \textit{if } b^2 \mid n \textit{ for some } b > 1 \\ (-1)^k & \textit{if } n \textit{ is the product of } k \textit{ distinct primes} \end{cases}$$

EXERCISE 14.4.   Find $\mu(n)$ for $n = 1, 2, 3, 4, 5, 6, 7, 8, 9$, and 10.

EXERCISE 14.5.   Show that $\mu(n)$ is a numerical function by showing that it assigns some definite value to each positive integer.

EXERCISE 14.6.   Compare $\mu(2)$, $\mu(3)$, and $\mu(6)$. Compare $\mu(9)$, $\mu(5)$, and $\mu(45)$.

EXERCISE 14.7.   Prove that $\mu(n)$ is a multiplicative function.

Next we are going to evaluate the Möbius function $\mu(d)$ for all the divisors d of an integer n and add them together.

EXERCISE 14.8.   Write the positive divisors d of 30, find μ of each, and add the results.

EXERCISE 14.9.   Find $\mu(d)$ for each positive divisor of 45. Add the μ values.

Theorem 14.1 *The sum $\sum_{d\mid n}\mu(d)$ is zero unless $n = 1$, in which case $\sum_{d\mid 1}\mu(d) = \mu(1) = 1$.*

Proof: Directly from the definition we have $\mu(1) = 1$. If $n > 1$, then n has a prime power factorization $n = p_1^{a_1} p_2^{a_2} \ldots p_s^{a_s}$, with $s > 0$ and each $a_i > 0$. The divisors d of n are all the integers of the form $p_1^{b_1} p_2^{b_2} \ldots p_s^{b_s}$, with b_i in the range $0 \leqslant b_i \leqslant a_i$, for each $i = 1, 2, \ldots, s$.

Each d for which one of the exponents b_i is greater than 1 has $\mu(d) = 0$, from the way the Möbius function is defined for such cases. Then the divisors that have non-zero values are

$$1, p_1, p_2, \ldots, p_s, p_1 p_2, p_1 p_3, \ldots, p_{s-1} p_s,$$
$$p_1 p_2 p_3, \ldots, \ldots, p_1 p_2 p_3 \cdots p_s.$$

For these, we have

$$\mu(1) = 1, \ \mu(p_1) = \mu(p_2) = \ ... \ = \mu(p_s) = -1,$$

$$\mu(p_1 p_2) = \mu(p_1 p_3) = \ ... \ = \mu(p_{s-1} p_s) = +1,$$

$$\mu(p_1 p_2 p_3) = \ ... \ = -1,$$

and so on.

From Definition 14.1 the μ value is -1 for each combination of the s primes taken an odd number at a time, $+1$ for each combination taken an even number at a time. If we expand the product $(1-p_1)(1-p_2)...(1-p_s)$, we get the sum: 1, minus the divisors d formed from the s primes taken one at a time, plus the products of two at a time, minus the products of three at a time, and so on. For instance, for $s = 3$, the product

$$(1-p_1)(1-p_2)(1-p_3)$$

$$= 1 - p_1 - p_2 - p_3 + p_1 p_2 + p_2 p_3 + p_1 p_3 - p_1 p_2 p_3$$

gives us the sum: 1, minus the divisors formed from one prime at a time, plus the divisors formed as products of two primes, minus the divisor formed from all three primes. More generally, we have

$$(1-p_1)(1-p_2)...(1-p_s)$$

$$= 1 - p_1 - p_2 ... - p_s + p_1 p_2 + p_1 p_3 + \ ... \ + p_{s-1} p_s$$

$$- p_1 p_2 p_3 ... + (-1)^s p_1 p_2 ... p_s. \tag{1}$$

If we replace each of the primes in the expanded sum (1) by $+1$, the result is precisely the sum of the $\mu(d)$ values, that is, $\mu(d) = -1$ for divisors that are products of an odd number of primes, and $\mu(d) = +1$ for products of an even number of primes, and $\mu(d) = 1$ for $d = 1$. We can illustrate this for $s = 3$:

$$1 - 1 - 1 - 1 + 1 + 1 + 1 - 1 = 0.$$

Then $\sum_{d|n} \mu(d)$ equals the sum (1) with each prime replaced by 1. From the left member of formula (1), with the primes replaced by 1, we have

$$\sum_{d|n} \mu(d) = (1-1)(1-1)(1-1)...(1-1) = (1-1)^s = 0^s = 0. \ \blacksquare$$

EXERCISE 14.10.   Calculate $\sum_{d|24} \mu(d)$.

EXERCISE 14.11.   Calculate $\sum_{d|35} \mu(d)$.

14.2 THE MÖBIUS INVERSION FORMULA

A. F. Möbius (1790–1866) was a student of Gauss. His name may already be familiar to you because of his famous one-sided surface, the "Möbius strip", shown in Figure 14.1.

To make a Möbius strip, make one half-twist in a paper band before joining the ends. A fly can walk the whole strip without going over an edge, as it has only one side. For a further experiment, try cutting lengthwise along the center of the strip.

Figure 14.1   A Möbius strip.

The Möbius inversion formula allows us to turn a certain kind of numerical function $F(n)$ inside out. Let $F(n)$ be defined as

$$F(n) = \sum_{d|n} f(d);$$

that is, let the function $F(n)$ be defined in terms of some other numerical function, $f(d)$, on the divisors of n. Our object is to learn about the function f from the function F that depends on it. This is somewhat like having a function y defined as $3x+7$ and "solving" for x in terms of y: $x = (y-7)/3$.

The functions $\tau(n)$ and $\sigma(n)$ of Chapter 12 are examples of such sum functions $F(n)$. Take $f(d) = 1$. Then $F(n) = \sum_{d|n} f(d) = \sum_{d|n} 1 = \tau(n)$, the number of divisors of n. Now take $f(d) = d$. Then $F(n) = \sum_{d|n} f(d) = \sum_{d|n} d = \sigma(n)$, the sum of the divisors of n.

Theorem 14.2 [The Möbius Inversion Formula]   *Let $F(n) = \sum_{d|n} f(d)$ be a numerical function. Then $f(n) = \sum_{d|n} \mu(d) F(n/d)$.*

Proof: We are going to show that the formula $\sum_{d|n} \mu(d) F(n/d)$ reduces to $f(n)$, where $F(n)$ is defined by hypothesis to be $\sum_{d|n} f(d)$. Since $F(n)$ is a numerical function, it assigns a value to each positive integer, including the integer n/d. We have

$$\sum_{d|n} \mu(d) F(n/d) = \sum_{d|n} \mu(d) \sum_{d'|n/d} f(d').$$

As in the special case shown in Figure 14.2, we rearrange this double sum so as to take advantage of Theorem 14.1. Let k be a divisor of n. Then

---

For $n = 6$,

$$\sum_{d|6} \mu(d) \sum_{d'|6/d} f(d')$$

$$= \mu(1)[f(1)+f(2)+f(3)+f(6)] + \mu(2)[f(1)+f(3)]$$
$$+ \mu(3)[f(1)+f(2)] + \mu(6)f(1)$$
$$= f(1)[\mu(1)+\mu(2)+\mu(3)+\mu(6)] + f(2)[\mu(1)+\mu(3)]$$
$$+ f(3)[\mu(1)+\mu(2)] + f(6)\mu(1)$$
$$= f(1)\cdot 0 + f(2)\cdot 0 + f(3)\cdot 0 + f(6)\cdot 1 = f(6)$$

Figure 14.2   Rearrangement of a double sum.

---

the terms of the double sum that contain $f(k)$ are the terms with $d' = k$, so that $k\,|\,(n/d)$, or $n/d = kq$, for some quotient q. The multipliers $\mu(j)$ for $f(k)$, then, have $n/j = kq$, or $jq = n/k$, so that $j\,|\,(n/k)$. Lumping together the terms in $f(k)$,

$$\sum_{d|n} \mu(d) \sum_{d'|n/d} f(d') = \sum_{k|n} f(k) \sum_{j|n/k} \mu(j).$$

But from Theorem 14.1, $\sum_{j|n/k} \mu(j)$ is zero unless $n/k = 1$, so the sum reduces to just that term

$$\sum_{\substack{k|n \text{ and} \\ n/k=1}} f(k)\cdot 1 = f(n). \quad \blacksquare$$

The converse to this theorem also holds, by almost the same proof; in fact, by careful definition of functions one can turn the function inside out again by Theorem 14.2.

Theorem 14.3   If $f(n) = \sum_{d|n} \mu(d) F(n/d)$, then $F(n) = \sum_{d|n} f(d)$.

Proof:

$$\sum_{d|n} f(d) = \sum_{d|n} \sum_{d'|d} \mu(d') F(d/d') = \sum_{k|n} F(k) \sum_{d'|n/k} \mu(d') = \sum_{\substack{k|n \text{ and} \\ n/k=1}} F(k)\cdot 1 = F(n). \quad \blacksquare$$

EXERCISE 14.12.   Use Theorem 14.2 to invert $\tau(n) = \sum_{d|n} 1$. Verify that for $n = 20$, $1 = \sum_{d|n} \mu(d)\tau(n/d)$.

EXERCISE 14.13.   Use Theorem 14.2 to invert $\sigma(n) = \sum_{d|n} d$, and verify that for $n = 20$, $n = \sum_{d|n} \mu(d)\sigma(n/d)$.

The inversion formula can be used to check several of the relations we have already discovered by some complicated counting. These applications, as you will see, amount to the kind of licensed trickery mathematicians term "elegant". You may have noticed a scarcity of applications to physics or engineering or even a lack of strong cohesiveness to a central idea in number theory. However, in number theory you have a chance to see at an elementary level the aesthetic side of mathematics, pursued almost as an art form for its own sake.

In Chapter 7 we counted the $\phi(m)$ integers less than or equal to m and prime to m, and then in Exercises 8.6 and 8.7 we proved that $m = \sum_{d|m} \phi(d)$. As our first application of the Möbius inversion formula let us deduce the result $m = \sum_{d|m} \phi(d)$ by means of the converse Theorem 14.3. To find $\phi(m)$, start with all m integers i in the range $1 \leqslant i \leqslant m$. Subtract the number that are multiples of a prime divisor p_1 of m; that is, subtract m/p_1 for the integers $p_1, 2p_1, \ldots, (m/p_1)p_1$ that are not prime to m. Also subtract the number m/p_2 that are multiples of a distinct prime that divides m, subtract m/p_3, and so on for each of the distinct prime divisors of m. Now notice that each integer that is a multiple of the product $p_1 p_2$ has been subtracted twice, once as a multiple of p_1 and again as a multiple of p_2. For this reason we add the number of multiples of $p_1 p_2$, of $p_1 p_3$, and so on, taking the prime divisors of m two at a time. Then, noticing that we have added twice for multiples of three distinct primes, we subtract the number of integers i that are such multiples, and so on, until we have a formula

$$\phi(m) = m - \sum_{\substack{p_1|m \\ \text{for all} \\ \text{single prime} \\ \text{divisors}}} m/p_1 + \sum_{\substack{p_1p_2|m \\ \text{all prime} \\ \text{divisors taken} \\ \text{2 at a time}}} m/p_1 p_2 - \sum_{\substack{p_1p_2p_3|m \\ \text{all prime} \\ \text{divisors taken} \\ \text{3 at a time}}} m/p_1 p_2 p_3$$

$$+ \cdots + (-1)^s \sum_{p_1p_2\ldots p_s|m} m/p_1 p_2 \ldots p^s.$$

With the Möbius function, we can simplify this to

$$\phi(m) = \sum_{d|m} \mu(d) m/d.$$

Then by Theorem 14.3, with $F(m/d) = m/d$, we have $F(m) = m = \sum_{d|m} \phi(d)$.

As a second application of the Möbius inversion formula, we apply Theorem 14.2 to show that $\phi(n)$ is a multiplicative function. We subject the formula

$$m = \sum_{d|m} \phi(d)$$

to the inversion of Theorem 14.2, with $f(d) = \phi(d)$ and $F(m) = m$, to deduce that

$$f(m) = \phi(m) = \sum_{d|m} \mu(d) F(m/d) = \sum_{d|m} \mu(d) \cdot m/d.$$

Suppose that m and n are relatively prime integers. Then since $(m, n) = 1$, each divisor $d > 1$ of mn can be written as $d = d_1 d_2$, where $d_1 \mid m$ and $f_2 \mid n$ but $(d_1, n) = 1$ and $(d_2, m) = 1$. Since the Möbius function is multiplicative we have

$$\mu(d_1 d_2) = \mu(d_1) \mu(d_2).$$

Then

$$\phi(mn) = \sum_{d \mid mn} \mu(d) \frac{mn}{d}$$

$$= \sum_{d_1 d_2 \mid mn} \mu(d_1 d_2) \frac{mn}{d_1 d_2}$$

$$= \sum_{d_1 \mid m} \sum_{d_2 \mid n} \mu(d_1) \mu(d_2) (m/d_1)(n/d_2)$$

$$= \sum_{d_1 \mid m} \mu(d_1) m/d_1 \sum_{d_2 \mid n} \mu(d_2) n/d_2$$

$$= \phi(m) \phi(n).$$

Thus we have proved that the ϕ-function is multiplicative, using the fact that the Möbius function is multiplicative.

As a third application, we evaluate $\phi(m)$ for a prime power. Suppose that m is a power of a prime, $m = p^w$. Since $\mu(p^2) = \mu(p^3) = \cdots = 0$ for any power above one, we have

$$\phi(p^w) = \sum_{d \mid p^w} \mu(d) p^w/d$$

$$= \mu(1) \cdot p^w + \mu(p) p^w/p$$

$$= 1 \cdot p^w + (-1) p^{w-1}$$

$$= p^w - p^{w-1}$$

$$= p^{w-1}(p-1),$$

which agrees with our result in Theorem 7.2.

CHAPTER 15

Fermat's Last Theorem

In our discussion of Bachet in Chapter 9 we mentioned his 1621 edition of *Diophantus* and the fact that a copy of it became a notebook for Fermat. One of the problems in Bachet's *Diophantus* was to express a square as the sum of two squares:

$$z^2 = x^2 + y^2.$$

In the margin Fermat wrote, "However, it is impossible to write a cube as the sum of two cubes, a fourth power as the sum of two fourth powers and in general any power beyond the second as the sum of two similar powers. For this I have discovered a truly wonderful proof, but the margin is too small to contain it."

This statement has become known as "Fermat's Last Theorem," and it is "last" if only in the sense that it is the last of Fermat's many statements to receive satisfactory proof or counter-example. We still do not know whether it is indeed so that

$$z^n = x^n + y^n$$

can have no solutions in non-zero integers for $n > 2$.

Unless somebody eventually shows the statement to be false, we can never know definitely whether Pierre de Fermat (1601–1665), lawyer and gentleman-scholar, had a correct proof. Mathematics was Fermat's favorite among his many intellectual pursuits. He carried on a vigorous correspondence with other scholars throughout Europe, proposing problems and ingenious solutions. Of all his mathematical contributions, including an early version of calculus, his number theory achievements seem to have brought him the most pleasure. So far as the excellence of his proofs goes, Fermat might well have proved his statement.

However, this elusive question has fooled some real experts. Among the baffled were the great brains of Cauchy, Gauss, Euler, Abel, Kummer,

Dirichlet, and many more. In fact, mathematicians have become so wary now that they still fail to accept the statement, although it has been proved for over 4000 individual cases $n = 3$, $n = 4$, $n = 5$, and so on.

Cauchy was one of the many who announced a "proof" of Fermat's Last Theorem, but his proof rested on the assumption that a unique factorization principle like the one for integers (Theorem 4.1) holds also for various extensions of the integers, called "algebraic integers." After a confident start, reinforced by early successes, mathematicians were suddenly brought up short by a counter-example, for $p = 23$. For all smaller primes unique factorization held, just as for the integers, but for $p = 23$ factorization into "primes" was not unique. After this discovery Cauchy published lots of mathematics in various fields, but no more papers in number theory!

In Chapter 2 we introduced the *ideal* (a, b) of linear combinations $ax + by$. Ideals were an extension of the "ideal numbers" invented by Kummer. Kummer introduced ideal numbers to generalize the notion of greatest common divisor for the purpose of attacking Fermat's Last Theorem.

Congruences have entered in, also, as some mathematicians, including Sophie Germain, a correspondence student of Gauss, have investigated the companion problem:

Prove that, for p a prime greater than 2,

$$z^p = x^p + y^p$$

can have no solutions not congruent to zero (mod p).

So we leave you with an unsolved problem, but one that has inspired quantities of number theory, one that in content as well as in history ties together much of what you have been learning.

ANSWERS TO SELECTED EXERCISES

CHAPTER 2

2.1, p. 9.   For 10 players, $52 = 10 \cdot 5 + 2$, for example.

2.2, p. 10.   4000, for example

2.4, p. 11. $17 = 6 \cdot 2 + 5$; $q = 2$, $r = 5$

2.9, p. 12. $2 \cdot 3 = 6$, $2 \cdot 4 = 8$, so there is no integer q for which $2 \cdot q = 7$. The non-integral rational (fractional) number $7/2$ satisfies

$$2 \cdot \frac{7}{2} = 7.$$

2.11, p. 12.   Yes, the quotient q can be negative or zero.

2.14, p. 13.   Try 5, for instance, or 19.

2.15, p. 13. $6i$, where i is an integer

2.20, p. 13.   Since $a \mid b$, there is a q_1 for which $b = aq_1$; since $b \mid x$, there is a q_2 for which $x = bq_2$. Then $x = bq_2 = (aq_1)q_2 = a(q_1 q_2)$. Then $a \mid x$, with quotient $q_1 q_2$.

2.22, p. 14.   Note that zero is excluded in Definition 2.1.

2.26, p. 17. $r = -2$, $s = 1$

2.29, p. 17. $661 = 341 \cdot 1 + 320$

$341 = 320 \cdot 1 + 21$

$320 = 21 \cdot 15 + 5$

$21 = 5 \cdot 4 + 1 \longleftarrow$

$5 = 1 \cdot 5 + 0$

Therefore,

$1 = 21 - 5 \cdot 4$

$\quad = 21 - (320 - 21 \cdot 15)4 = 21(1 + 15 \cdot 4) - 320 \cdot 4$

$\quad = (341 - 320 \cdot 1)(61) - 320 \cdot 4 = -320 \cdot 65 + 341 \cdot 61$

$\quad = -65(661 - 341 \cdot 1) + 341 \cdot 61$

$\quad = (-65)661 + (126)341$

CHAPTER 3

3.3, p. 22. $177 = 3 \cdot 59$; $148 = 2^2 \cdot 37$. $(177, 148) = 1$.

3.6. p. 23. $101 = \sqrt{101} \cdot \sqrt{101} = d \cdot \dfrac{101}{d}$. If $d > \sqrt{101}$, then $\dfrac{101}{d} < \sqrt{101}$.

CHAPTER 4

4.4, p. 30. $(a, b) = 29^7 \cdot 37$.

4.10, p. 31. $726 = 2 \cdot 3 \cdot 11^2$; $1431 = 3^3 \cdot 53$; $[726, 1431] = 2 \cdot 3^3 \cdot 11^2 \cdot 53$

4.14, p. 31. $11(2 \cdot 3 \cdot 11^2 \cdot 23) = 8349 \cdot 22$

CHAPTER 5

5.1, p. 33. $14 \equiv 95 \equiv -40 \equiv -4 \equiv 5 \pmod 9$

5.2, p. 33. $5 \equiv 11 \pmod 2$, $5 \equiv 11 \pmod 3$
               $5 \not\equiv 11 \pmod 4$, $5 \not\equiv 11 \pmod{100}$

5.10, p. 33.   Yes, for if $a = 6 + q10$, then $a = (3 + q5)2$.

5.17, p. 37. $1 \equiv -7$
               3
               4
               $5 \equiv -3$
               $7 \equiv -1 \equiv 15 \equiv -9$
               Three classes, 0, 2, and 6, are not represented.

5.23, p. 39. $\$3.17 + \$21.74 + \$18.98 = \$43.89$
               $\equiv 4 + 3 + 8 + 9 \equiv 6 \pmod 9$
               $\$3.17 \equiv 2$, $\$21.74 \equiv 5$, $\$18.98 \equiv 8$, and
               $2 + 5 + 8 \equiv 6 \pmod 9$

5.25, p. 40.   South receives the 44th card.

5.34, p. 44.   Written to base 12, an integer divisible by 12 ends in 0, an integer divisible by 3 ends in 0, 3, 6, or 9, an integer divisible by 4 ends in 0, 4, or 8, and the sum of the digits of an integer divisible by 11 is congruent to zero modulo 11.

CHAPTER 6

6.7, p. 50. $200x \equiv 10 \pmod{35}$. $(200, 35) = 5$.
               $40x \equiv 2 \pmod 7$

$$40x \equiv -2x \equiv 2 \pmod 7$$
$$x \equiv -1 \equiv 6 \pmod 7$$
Solutions $\pmod{35}$ are 6, 13, 20, 27, 34.

6.13, p. 50. $x \equiv 2 \pmod 7$ and $x \equiv 3 \pmod 5$.
$x \equiv 23 \pmod{35} \equiv 2 \pmod 7$ and $\equiv 3 \pmod 5$.

6.16, p. 51.   Since $(4,100) = 4$, there are years requiring the finer adjustments. (Actually, still finer ones are necessary.) 1900.

CHAPTER 7

7.2, p. 55.   Of the positive numbers less than or equal to 14,
$$1, \cancel{2}, 3, \cancel{4}, 5, \cancel{6}, \cancel{7}, \cancel{8}, 9, \cancel{10}, 11, \cancel{12}, 13, \cancel{14},$$
six are relatively prime to 14, so that $\phi(14) = 6$. Of the positive numbers less than or equal to 18, six are relatively prime to 18, so that $\phi(18) = 6$.

7.8, p. 58. $3^2 = 9 \equiv 4 \pmod 5$
$4^3 \equiv 4 \cdot 4^2 \equiv 4 \cdot 1 \equiv 4 \pmod 5$

7.11, p. 60. $C_i^{n+1} = \dfrac{(n+1)!}{i!(n+1-i)!} = \dfrac{(n+1)n!}{i!(n+1-i)(n-i)!}$

$$= \frac{(n+1-i)n!}{(n+1-i)i!(n-i)!} + \frac{i \cdot n!}{i(i-1)!(n+1-i)(n-i)!}$$

$$= \frac{n!}{i!(n-i)!} + \frac{n!}{(i-1)!(n-(i-1))!}$$

$$= C_i^n + C_{i-1}^n.$$

7.14, p. 63. $c^{15} - 1 = c^{3 \cdot 5} - 1 = (c^3 - 1)(c^{12} + c^9 + c^6 + c^3 + 1)$, and
$c^{15} - 1 = (c^5 - 1)(c^{10} + c^5 + 1)$, for example.

CHAPTER 8

8.1, p. 65. $2^4 = 16 \equiv 1 \pmod{15}$. $4 \mid \phi(15)$, since $\phi(15) = 8$. If $2^f \equiv 1 \pmod{15}$, then $f = 4q + r$, with $0 \leqslant r < 4$, so that
$$2^f = 2^{4q+r} = 2^{4q} \cdot 2^r = (2^4)^q \cdot 2^r \equiv 1^q \cdot 2^r \equiv 2^r \equiv 1,$$
$r = 0$, and $4 \mid f$.

8.8, p. 69. $1^4 \equiv 1$, $2^4 \equiv 5$, $3^4 \equiv 9^2 \equiv 4$, $4^4 \equiv 5^2 \equiv 3$, $5^4 \equiv 3^2 \equiv 9$.

CHAPTER 9

9.3, p. 74. $(+4)6 + (-2)(14) = -4.$
$(x_0 + b'n, y_0 - a'n) = (4 + 7n, -2 - 3n), n = 28.$
Sells 200 shares, buys 86 shares.

9.5, p. 74.   If $m = b$, then there are 30 guests, or 15 father-son pairs.

9.12, p. 76. $(5, 12, 13)$. $r = 3, s = 2$. $3^2 - 2^2 = 5, 2 \cdot 3 \cdot 2 = 12, 3^2 + 2^2 = 13$.

CHAPTER 11

11.3, p. 93. $\bar{d}_6/\bar{s}_6 = 208/120.$

11.5, p. 93. $\bar{d}_{n+1}^2 - 3\bar{s}_{n+1}^2 = (3\bar{s}_n + \bar{d}_n)^2 - 3(\bar{s}_n + \bar{d}_n)^2$
$= 9\bar{s}_n^2 + 6\bar{s}_n \bar{d}_n + \bar{d}_n^2 - 3\bar{s}_n^2 - 6\bar{s}_n \bar{d}_n - 3\bar{d}_n^2$
$= -2\bar{d}_n^2 + 6\bar{s}_n^2 = -2(\bar{d}_n^2 - 3\bar{s}_n^2)$

CHAPTER 12

12.5, p. 96.   25 has divisors 1, 5, 25.
$\sigma(25) = 1 + 5 + 25 = 31; \tau(25) = 3.$

12.7, p. 96.   For $n > 1$, $\sigma(n) = \sigma^-(n) + n.$

12.9, p. 96. $\sigma(3^4) = 1 + 3 + 9 + 27 + 81 = 121; \tau(3^4) = 5.$

12.13, p. 98. $\sigma(5) = 6, \sigma(7) = 8, \sigma(35) = 1 + 5 + 7 + 35 = 48 = 6 \cdot 8$

12.15, p. 98. $496 = 2^4(2^5 - 1)$. $2^5 - 1 = 31$ is a prime.

CHAPTER 13

13.3, p. 108.   1, 5, 12, 22, 35, 51, 70

13.9, p. 109. $8(k(k+1)/2) + 1 = 4k^2 + 4k + 1 = (2k+1)^2$

13.11, p. 109. $36 = p_3^8 = p_4^6 = p_{13}^3 = p_{36}^2.$

CHAPTER 14

14.3, p. 112. $f(1) = 1, f(10) = 55, f(20) = 210.$

14.4, p. 113. $\mu(1) = 1$, $\mu(2) = -1$, $\mu(3) = -1$, $\mu(4) = 0$, $\mu(5) = -1$,
$\mu(6) = 1, \mu(7) = -1, \mu(8) = 0, \mu(9) = 0, \mu(1) = 1.$

14.9, p. 113. $\mu(1) + \mu(3) + \mu(5) + \mu(9) + \mu(15) + \mu(45)$
$$= 1 + (-1) + (-1) + 0 + 1 + 0 = 0.$$

14.11, p. 114. $\sum_{d|35} \mu(d) = \mu(1) + \mu(5) + \mu(7) + \mu(35)$
$$= 1 - 1 - 1 + 1 = 0.$$

14.13, p. 116. $\sum_{d|20} \mu(d)\sigma(20/d)$
$$= \mu(1)\sigma(20) + \mu(2)\sigma(10) + \mu(4)\sigma(5) + \mu(5)\sigma(4)$$
$$+ \mu(10)\sigma(2) + \mu(20)\sigma(1)$$
$$= 1 \cdot 42 + (-1)18 + 0 \cdot 6 + (-1)7 + 1 \cdot 3 + 0 \cdot 1$$
$$= 20.$$

Index

Page on which definition appears is distinguished by **boldface** type.

SYMBOLS, in order of appearance